Lecture Notes in Economics and Mathematical Systems

678

More information about this series at
http://www.springer.com/series/300

Moritz Mühlenthaler

Fairness in Academic Course Timetabling

Springer

Moritz Mühlenthaler
Dept. of Computer Science 12
University of Erlangen-Nuremberg
Erlangen
Germany

Dissertation submitted in partial fulfilment of the requirements for the degree of Doctor of Engineering (Dr.-Ing.) at the University of Erlangen-Nuremberg, 2014

ISSN 0075-8442 ISSN 2196-9957 (electronic)
ISBN 978-3-319-12798-9 ISBN 978-3-319-12799-6 (eBook)
DOI 10.1007/978-3-319-12799-6
Springer Cham Heidelberg New York Dordrecht London

Library of Congress Control Number: 2015930113

Printed on acid-free paper

Springer is part of Springer Science+Business Media (www.springer.com)

Abstract

Creating timetables is a complex and recurring task at a university. In this thesis, we will deal with formal problems related to course timetabling, issues concerning the formalization, and solution approaches.

First, we study the University Course Timetabling Problem (UCTP), a very general formalization of the course timetabling task in terms of a search problem. The task is to find a feasible timetable, i.e., an assignment of lectures to rooms and timeslots such that no room is occupied by two lectures at the same time and no two lectures are taught simultaneously if they are attended by the same students. This problem captures the essence of various more specific course timetabling problems. We investigate the structure of the UCTP search space and establish sufficient conditions for the connectedness of clash-free timetables under the Kempe-exchange operation. Furthermore, we propose and evaluate the Kempe Insertion Heuristic (KIH), a heuristic algorithm for solving the UCTP.

Second, we consider fairness aspects of course timetabling with a particular focus on how fairness can be formalized and measured. We introduce fairness objectives based on Jain's fairness index and max–min fairness to a popular benchmarking problem model. We show that a problem decomposition which isolates room and timeslot assignment has similar computational properties to an analogous decomposition of the standard benchmarking problem. Additionally, we present several approaches to measuring fairness differences, and we propose and evaluate heuristic approaches to solving the benchmarking problems with fairness objectives.

Finally, we present a comprehensive case study on our implementation of a course timetabling system at the Friedrich-Alexander-Universität Erlangen-Nürnberg (FAU) school of engineering. The system has been used in practice since 2011. We report on our problem model the organization of the course timetabling process, as well as the reception by students and lecturers.

Acknowledgments

I would like to express my sincere gratitude to Prof. Dr. Rolf Wanka for his guidance and patience.

I would also like to thank everyone involved in the coordination of the timetabling process, Sonja Gebhard, Michael Wensing, and of course the "Raumplanungsgruppe" of the first and second generation. Without your efforts and your support, the realization of the timetabling system would not have been possible. In particular, I would like to thank Prof. Dr.-Ing. Johannes Huber for initiating the automated course timetabling project.

Last but not least, I am indebted to {Moritz Schmid, Manuel Schmitt} (in no particular order) for the discussions, the feedback, and the good time.

Contents

List of Symbols

$\preceq$	Max–min fairness relation
$\prec$	Strict max–min fairness relation
$\preceq_{\mathrm{P}}$	Weak Pareto dominance relation
$\prec_{\mathrm{P}}$	Strict Pareto dominance relation
α	Resource availability function
β	Period preference function
γ	Room preference function
δ	Minimum vertex degree of a graph
Δ	Maximum vertex degree of a graph
κ	Kempe-exchange
$\mathbb{N}$	Natural numbers
ρ	Assignment of events to rooms
σ	Vertex ordering
τ	Timetable
φ	Boolean formula
χ	Chromatic number of a graph
C	Set of classes
$\mathcal{A}$	Variables of a boolean formula
C	Set of courses
L	Course conflicts
F	Room conflicts
deg	Degeneracy of a graph
diam	Diameter of a graph
d	Degree of a vertex
$\mathcal{D}$	Problem domain
E	Set of events
girth	Girth of a graph
h	Period restriction number
$\mathcal{I}$	Instance of a combinatorial problem
NP	Complexity class NPTIME

N	Neighborhood of a vertex
PSPACE	Complexity class PSPACE
P	The complexity class PTIME
P	A set of timeslots
R	Set of rooms
$\mathcal{S}$	Set of solutions
size	Size of a CNF formula
T	Set of teachers
U	Set of curricula
var	Number of variables in a CNF formula
wp	Welsh-Powell number of a graph

Acronyms

ACO	Ant Colony Optimization
ACP	Algorithm Configuration Problem
AMOSA	Archived Multi-Objective Simulated Annealing
BACP	Balanced Academic Curriculum Problem
BFS	Breadth-first Search
BSSP	Basic Student Sectioning Problem
CB-CTT	Curriculum-based Course Timetabling
CNF	Conjunctive Normal Form
DPLL	Davis–Putnam–Logemann–Loveland
EA	Evolutionary Algorithm
EIP	Event Insertion Problem
ETP	Examination Timetabling Problem
ETT	Examination Timetabling
FAU	Friedrich-Alexander-Universität Erlangen-Nürnberg
GA	Genetic Algorithms
GBACP	Generalized Balanced Academic Curriculum Problem
GD	Great Deluge
GGA	Grouping Genetic Algorithm
GLBOP	Generalized Lexicographic Bottleneck Optimization Problem
HC	Hill Climbing
HGHH	Hybrid Graph-based Hyper Heuristic
ILP	Integer Linear Programming
ILS	Iterated Local Search
ITC2007	International Timetabling Competition 2007
JFI-CB-CTT	Jain's Fairness Index Curriculum-based Course Timetabling
KIH	Kempe Insertion Heuristic
KX	Kempe-exchange Neighborhood
LBAP	Lexicographic Bottleneck Assignment Problem
LBFS	Lexicographic Breadth-first Search
LBOP	Lexicographic Bottleneck Optimization Problem
LCD	Largest Color Degree

LD	Largest Degree
LSAP	Linear Sum Assignment Problem
LSD	Least Saturation Degree
LS	Local Search
SLS	Stochastic Local Search
LVOP	Lexicographic Vector Optimization Problem
LWD	Largest Weighted Degree
ME	Move Event Neighborhood
MMF-CB-CTT	Max–min Fair Curriculum-based Course Timetabling
NOLH	Nearly Orthogonal Latin Hypercubes
PE-CTT	Post-enrollment Course Timetabling
PKC	Pair-wise Kempe Chain
PSO	Particle Swarm Optimization
RO	Random Ordering
SA	Simulated Annealing
SAT	Boolean satisfiability
SE	Swap Event Neighborhood
SOP	Min-sum Optimization Problem
SO	Saturation Ordering
SSP	Student Sectioning Problem
STP	School Timetabling Problem
TF-CB-CTT	TechFak Curriculum-based Course Timetabling
TS	Tabu Search
uar	uniformly at random
UCTP	University Course Timetabling Problem
VNS	Variable Neighborhood Search
WEO	Weak Elimination Ordering
WMW	Wilcoxon-Mann-Whitney

Chapter 1
Introduction

> *In regard to hedonism and utilitarianism, I believe that it is indeed necessary to replace their principle:* maximize pleasure! *by one which is probably more in keeping with the original views of Democritus and Epicurus, more modest, and much more urgent. I mean the rule:* minimize pain!
>
> Karl R. Popper

Creating timetables is a complex and recurring task at any university. Basically, the goal is to create a weekly schedule of the teaching activities by assigning lectures to rooms and time periods (*timeslots*). From the perspective of a computer scientist, it seems natural to formalize the task in terms of a computational problem so that timetables can be created in an automated fashion. In this work we will deal with formal computational problems related to course timetabling as well as the problem of formalizing the course timetabling task. One of our main topics occurring throughout this work will be the formalization of the *quality* of a timetable and how to measure it.

The research presented in this work has been accompanied by the development of a custom timetabling system for the school of engineering at the Friedrich-Alexander-Universität Erlangen-Nürnberg (FAU). The main motivations for formalizing and automating the timetabling task were an anticipated doubling of the number of freshmen as well as structural changes concerning the organization of study programs. The expected increase in the number of freshmen was largely due to two cohorts finishing the Bavarian secondary education simultaneously in 2011. The challenge was to implement a working course timetabling system within a time frame of roughly two and a half years. Apart from the computational aspects—timetable construction is generally a tough computational problem—considerable effort was required to embed the timetabling system in the social and institutional fabric.[1]

In our timetabling system, in a sense, we followed Popper's appeal above for a transformation towards minimizing the figurative "pain", or burden, of the students

[1] We invite the adventurous reader to have a peek at p. 123 for the visualization of a small part of the problem instance solved for the 2013 winter term.

M. Mühlenthaler, *Fairness in Academic Course Timetabling*, Lecture Notes in Economics and Mathematical Systems 678, DOI 10.1007/978-3-319-12799-6_1

and lecturers, to whom we will refer to as the *stakeholders*. One of the main concerns in this work is to take this idea one step further, seeking not only to minimize the pain in the (negative) utilitarian sense, but to establish *fairness* by considering the distribution of pain over the stakeholders. If this distribution is skewed, then a timetable may be perceived as unfair by the stakeholders, which undermines the acceptance of a timetabling system. As a consequence, we propose computational optimization problems that formalize the task of creating fair course timetables. In addition, we study the structure of these problems and provide and evaluate heuristic algorithms for them.

1.1 Overview

We will give a brief overview of this work and highlight the contributions of this work.

1.1.1 The University Course Timetabling Problem

The heart of the matter of the different timetabling problems studied in this work and a considerable part of the course timetabling literature is the following basic task: Find a timetable that satisfies the following two basic requirements: (1) no two lectures are taught in the same room at the same time, and (2) no two conflicting lectures are taught simultaneously. Two lectures are conflicting for example, if they are taught by the same teacher. This problem formulation is essentially equivalent to the one given in [dW85, Section 3.4] and will be referred to as the University Course Timetabling Problem (UCTP). For all relevant cases, the UCTP is a hard computational problem: the corresponding decision problem can be shown to be NP-complete by a straight-forward reduction from the graph coloring problem. Despite its simplicity, the UCTP is also of practical importance: For instance, the complex problem model currently used for creating course timetables at the FAU school of engineering is in essence an optimization version of the UCTP with some additional requirements.

In Chap. 2, we will review solution approaches to the UCTP and investigate combinatorial properties of the corresponding search space.

The Kempe Insertion Heuristic (KIH) The KIH is a randomized heuristic algorithm for solving the UCTP [MW10b*]. Similar to many existing approaches, the KIH tries to insert events in the timetable one-by-one without violating any of the requirements. The key idea is that an additional event can be inserted in the timetable by deliberately clearing timeslots from conflicting events by using Kempe-exchanges. We perform computational experiments in order to evaluate the KIH algorithm on publicly available, artificially generated, "hard" UCTP instances and compare its performance to other approaches from the literature and a SAT solver.

Search Space Analysis We will establish conditions that certify the connectedness of all timetables that satisfy the conflict requirements with respect to a local modification operation, the *Kempe-exchange.* For this purpose, we study the reconfiguration variant of the graph coloring problem. The connectedness of timetables is relevant for example for the class of *two-step* heuristic optimization algorithms, see for example [Lew06]. A two-step algorithm first solves the basic problem above and then searches among all timetables satisfying the above requirements one that is most suitable with respect to other institutional requirements. Furthermore, we establish sufficient conditions for the connectedness of timetables that also satisfy timeslot availability requirements. These requirements stipulate that lectures can only be taught in specific timeslots. We additionally investigate the reconfiguration variants of the graph coloring problem for particular graph classes. We prove that any coloring of a chordal (triangulated) graph with n nodes can be transformed into any other coloring using at most n Kempe-exchanges.

1.1.2 Fairness in Academic Course Timetabling

In Chap. 3 we will formalize the task of creating *fair* course timetables in terms of optimization problems. That is, the task is to find a timetable that is optimal with respect to some quality measure. The usual situation is that the quality of a timetable is determined by so-called soft constraints, which capture undesirable features of a timetable. Each violation of a soft constraint results in a penalty for the timetable, and the task is to find a timetable such that the sum of the penalties is minimal. However, this approach does not reflect how stakeholders judge the quality of a timetable. Each student or lecturer is mostly interested in the quality of his of her individual timetable. Situations may arise, in which a large part of penalty hits only a small group of the stakeholders, who would thus receive a poor timetable in comparison to others. Such timetables would likely be perceived as unfair by the stakeholders and may erode the acceptance of a timetabling system in practice. Our key idea is to consider the distribution of penalty over the stakeholders, with the goal of avoiding an unfair distribution of penalty. We present fair course timetabling problem formulations, a decomposition result, several approaches to measuring fairness differences, and solution approaches.

Problem Formulations We formalize the task of creating fair course timetables by giving two optimization problems which include fairness criteria in their objective functions. Both problems are derived from the popular Curriculum-based Course Timetabling (CB-CTT) problem used for benchmarking [DGMS07]. Our fairness criteria are based on *max-min fairness* and *Jain's fairness index* and the corresponding optimization problems are referred to as MMF-CB-CTT problem and JFI-CB-CTT problem [MW12*]. The two fairness notions have received considerable attention for example in the context of bandwidth allocation, and some modifications are necessary to adapt them to our setting.

Problem Decomposition The UCTP can be decomposed in a master and a subproblem by isolating the subtask of assigning the lectures in a single timeslot to rooms. This subtask can be solved efficiently for instance for the CB-CTT problem [LH10, LL12]. We establish an analogous decomposition result for the Max–min Fair Curriculum-based Course Timetabling (MMF-CB-CTT) problem [MW13*]. For this purpose, we introduce the Generalized Lexicographic Bottleneck Optimization Problem (GLBOP), which is an optimization problem asking for a max-min fair solution given finite multi-sets as penalty or utility values. We show that whenever the problem of minimizing the sum of the penalty values can be solved efficiently, then the corresponding GLBOP can also be solved efficiently.

Measuring Max-Min Fairness Max-min fairness is a quantitative fairness concept. That is, among two resource allocations, max-min fairness tells us which one is fairer, but not by how much. We introduce several approaches for measuring max-min fairness for two purposes: First, measuring the quality difference between two solutions is for instance a prerequisite for using the Simulated Annealing (SA) [KGV83] in order to solve problems with a max-min fairness objective. We propose three different approaches for measuring this quality difference and evaluate them using SA on standard timetabling benchmark instances [MW12*]. Second, due to the qualitative nature of max-min fairness, an evaluation of randomized heuristic algorithms for problems with a max-min fairness objective is restricted to methods based on ranking. In order to partially overcome this limitation, we propose an invertible mapping from the penalty/utility distributions to the natural numbers which preserves the max-min fairness ordering [MW13*]. This allows us, for example, to approximately compute an average penalty distribution produced by a randomized algorithm.

1.1.3 Real-World Academic Course Timetabling

In the fourth chapter we report on practical aspects of the course timetabling system which has been implemented at the FAU school of engineering. This system has been used successfully since the 2011 summer term and has enabled the school of engineering to cope with increasing student numbers and various structural changes. In this chapter, we focus on the organizational challenges related to the course timetabling task. We report on the process of creating timetables before and after the introduction of the timetabling system as well as the formal problem model, which has been proven in practice. Naturally, the organizational aspects of timetabling are very specific to the institution. Nevertheless, some of the experiences reported in Chap. 4 are likely to be relevant for other institutions as well and thus may yield a more general benefit.

Timetabling Process We present a structured approach to organizing the course timetabling in a curriculum-driven setting, similar to the work of Carter, who describes the course timetabling process for a demand-driven setting at the University of Waterloo [Car01]. The take-home message is that solving the computational problem is just one subtask, and ensuring the validity of the instances is at least as important as generating high quality timetables in the model.

Problem Model We specify the problem model which is currently used at the FAU school of engineering. The model is based on the CB-CTT benchmarking problem. We show in detail which additional requirements were considered in order to adapt the CB-CTT problem to our real-world setting.

Reception and Feedback We will present and discuss the feedback given by students and lecturers on the perceived quality of their timetables. The student feedback has been obtained through structured annual student satisfaction surveys. We can conclude from this feedback which aspects of a timetable are particularly undesirable for students, based on a large number of samples.

1.2 Computational Problems

We will give formal definitions of the computational problems studied in this work. We consider three general categories of problems: *search*, *reconfiguration*, and *optimization* problems. Here, we define these types of problems on an abstract level, so can skip the technical details later on. The general style of the presentation is inspired by Wanka [Wan06].

Informally, a computational decision problem can be stated in terms of a YES-NO question. In our application, answering such questions is typically not sufficient, although we will give them some consideration in the guise of reconfiguration problems. Rather, we are interested in problems that require from us to find a solution of some form, known as *search problems*. More formally, a search problem is defined as follows:

Definition 1.1 (Search Problem) A search problem Π is given by:

- $\mathcal{D}$: A set of inputs, referred to as problem instances.
- $\mathcal{S}(\mathcal{I})$, $\mathcal{I} \in \mathcal{D}$: The set of solutions to a problem instance $\mathcal{I}$, referred to as *search space*.

The task is to find a solution $\sigma \in \mathcal{S}(\mathcal{I})$ to a given instance $\mathcal{I}$.

Sometimes we distinguish between *solutions* and *feasible solutions*. In these cases, we can assume that $\mathcal{S}(\mathcal{I})$ are the feasible solutions.

Reconfiguration problems formalize the question, if a solution can be transformed into another solution in a step-by-step manner by some operation, such that after each step we have a valid solution. The given operation induces a symmetric adjacency relation on the solutions of a problem. Thus, we can equivalently ask if two solutions are adjacent with respect to this relation. Such problems have been popularized for example by the work of Ito et al. [IDH+11] and Cereceda et al. [CvdHJ08]. We will consider reconfiguration problems in order to obtain results on the structure of the search space of timetabling problem instances.

Definition 1.2 (Reconfiguration Problem) A reconfiguration problem is given by:

- A search problem, i.e., a domain $\mathcal{D}$ and solutions $\mathcal{S}(\mathcal{I}), \mathcal{I} \in \mathcal{D}$.
- Symmetric adjacency relations $\sim_{\mathcal{I}}$ on $\mathcal{S}(\mathcal{I}), \mathcal{I} \in \mathcal{D}$.

Given an instance $\mathcal{I} \in \mathcal{D}$ and two solutions $\sigma_1, \sigma_2 \in \mathcal{S}(\mathcal{I})$, the task is to decide if $\sigma_1 \sim_{\mathcal{I}}^{+} \sigma_2$, where $\sim_{\mathcal{I}}^{+}$ is the transitive closure of $\sim_{\mathcal{I}}$.

When creating timetables, we typically ask for one that is "best" according to some criterion. Such tasks are captured by optimization problems.

Definition 1.3 (Optimization Problem) An optimization problem is given by:

- A search problem, i.e., a domain $\mathcal{D}$ and solutions $\mathcal{S}(\mathcal{I}), \mathcal{I} \in \mathcal{D}$.
- *Objective functions* $f_{\mathcal{I}} : \mathcal{S}(\mathcal{I}) \to \mathbb{N}, \mathcal{I} \in \mathcal{S}(\mathcal{I})$.
- A goal $\in \{\min, \max\}$.

The task is to find for a given instance $\mathcal{I} \in \mathcal{D}$ a solution $\sigma^* \in \mathcal{S}(\mathcal{I})$ such that

$$f_{\mathcal{I}}(\sigma^*) = \text{goal}\,\{f_{\mathcal{I}}(\sigma) \mid \sigma \in \mathcal{S}(\mathcal{I})\} \ .$$

We can observe that reconfiguration and optimization problems as defined above are based on a search problem. We refer to this problem as the *underlying search problem*. For the sake of succinctness, we will just define the instance data and the task when we state specific computational problems later on. The domain and the solutions should be clear from the context if they are not given explicitly. In a slight abuse of notation, we will omit the index of the adjacency relation of a reconfiguration problem because it is merely a technical detail and refer to *the* adjacency relation. Similarly, we will omit the index of an objective function and refer to *the* objective function of an optimization problem.

1.3 Computational Experiments

In Chaps. 2 and 3, we will present the results of several computational experiments, which were conducted in order to evaluate randomized heuristic algorithms for timetabling problems. Here, we will briefly discuss the methods we employ. The methodology is more or less standard in the field.

In all our experiments, we adhere to the following generic procedure outlined in [BGK+95]:

1. Define the goals of the experiment

2. Choose measures of performance and factors to explore
3. Design and execute the experiment
4. Analyze the data and draw conclusions
5. Report the experiment's results

For our data analysis, we use basic tools from descriptive statistics (e.g., the mean and standard deviation of measured values), and statistical inference (hypothesis testing). A statistical hypothesis test allows us to reason about measured data by indicating if a result is likely to have occurred by coincidence.

Statistical Hypothesis Testing

We will just give some basic definitions related to hypothesis testing that will be useful for interpreting the results of our experiments. For a more in-depth treatment please refer to a textbook on statistical hypothesis testing, for example [Con06]. A test typically involves two hypotheses about an experiment, the null-hypothesis and the alternative hypothesis. The test decides for a given significance level α, if an observation provides sufficient evidence to reject the null-hypothesis in favor of the alternative. Lower values of α require stronger evidence for rejecting the null-hypothesis. Based on observed data, a statistical test computes the p-value, which is the probability of obtaining an observation at least as extreme as the given observation, under the assumption that the null-hypothesis is true. There is sufficient evidence to reject the null-hypothesis if the p-value is less than α. Typical values of α are between 0.05 and 0.01. In all our tests we will set $\alpha = 0.01$.

For our comparison of the performance of different algorithms, we will use the Wilcoxon-Mann-Whitney (WMW) test [Wil45]. The WMW test is a non-parametric rank-based test, that is, the decision if there is enough evidence to reject the null-hypothesis is based on the *ranks* of a number of measurements instead of the measured values. Please note that there are some peculiarities with different implementations of the WMW test in various statistics packages [BLS00]. We use implementation from the `coin` package [HvdWZ06] of the R statistics software [R D08] in our test, which behaves correctly with respect to the concerns raised in [BLS00]. The error rate accumulates if more than one test is performed: For example, in our experiments described in Sect. 3.6.2, we compare different approaches to quantifying fairness differences and for this purpose we perform 480 WMW tests in total. In each statistical test, there is a chance of falsely rejecting the null-hypothesis, referred to as the type I error rate. When performing multiple tests, the type I error rate accumulates over the tests and results in the *family-wise error rate*, which can be surprisingly high. For instance, assuming a type I error rate of 0.01, the probability of making at least one type I error in 480 WMW tests amounts to $1-(1-0.01)^{480} \approx 0.992$. This error accumulation needs to be kept in mind when interpreting the results of our experiments. From the methodological point of view, there seems to be no satisfactory cure, since the problem is inherent in the iterated use of statistical tests.

1.4 Graph-Theoretic Preliminaries

Timetabling problems and many of their properties can be conveniently be described in the language of graph theory. Here, we will give definitions of basic graph-theoretic terms which will be used throughout this work. For a detailed discussion of these basic terms and more advanced topics in graph theory we recommend [Die12].

Basic Definitions A graph $G = (V(G), E(G))$, for short $G = (V, E)$, consists of a set of *vertices*, or *nodes*, V and a set of *edges* $E \subseteq \{\{u, v\} \mid u, v \in V\}$. Unless stated otherwise, we assume that graphs are loopless and finite. If $u, v \in V$ and u is adjacent to v we may write $u - v$ whenever it is clear from the context which graph we refer to. The *neighborhood* $\mathrm{N}(v)$ of a vertex $v \in V$ is the set of vertices adjacent to it, i. e., $\mathrm{N}(v) = \{u \in V \mid u - v\}$. A graph H is a subgraph of G, denoted by $H \subseteq G$, if $V(H) \subseteq V(G)$ and $E(H) \subseteq E(G)$. The graph $G[U]$ is subgraph of G induced by the vertices $U \subseteq V(G)$, that is, $G[U] = (U, E')$, where $E' = \{\{u, v\} \in E \mid u, v \in U\}$. For convenience we may write $G[u_1, \dots, u_k]$ for the subgraph induced by the vertices $\{u_1, \dots, u_k\}$. $G - v$ is the graph which results from removing the vertex $v \in V(G)$ and all edges incident to v from G. The *complement* of G is the graph $\overline{G} = (V(G), \overline{E(G)})$, where $\overline{E(G)} = \{\{u, v\} \mid u, v \in V(G), \{u, v\} \notin E(G)$. K_n denotes the clique on n vertices. The graph K_6 is shown in Fig. 1.1a. The *degree* $d(v, G)$ of a vertex v in G is the number of vertices of G adjacent to v. A graph is k-regular, if each vertex has degree k. The *minimum degree* of G is $\delta(G) = \min_{v \in V(G)} d(v, G)$ and the *maximum degree* of G is $\Delta(G) = \max_{v \in V(G)} d(v, G)$. The *distance* of two nodes $u, v \in V(G)$ is the number of edges of a shortest path between u and v. The *diameter* $\mathrm{diam}(G)$ is the maximum distance of any two nodes of G. The *girth* of G is the minimum length of a cycle in G.

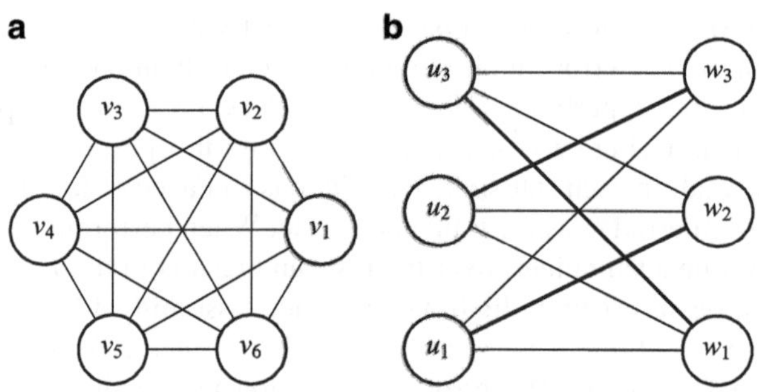

Fig. 1.1 Two example graphs. (**a**) The graph K_6. (**b**) The graph $K_{3,3}$

Vertex Coloring A *(vertex-)k-coloring* of a graph G is a mapping $c : V \to \{1, \ldots, k\}$ that assigns one of the colors $\{1, \ldots, k\}$ to each vertex of G. A coloring is called *proper*, if no two adjacent nodes have the same color. Unless stated otherwise, we will use the term *coloring* as a shorthand for a *proper coloring*. The least number k for which G admits a k-coloring is the *chromatic number* $\chi(G)$. A graph is *bipartite* if and only if it admits a 2-coloring. A k-coloring c of G decomposes the vertices of G into k independent sets called color classes. A set U of vertices is independent if $G[U]$ has no edges. A color class a contains all vertices of color a. We denote by $G(a,b)$ the bipartite subgraph induced by the color classes a and b. A connected component in $G(a,b)$ is referred to as *Kempe-component*. As an example, consider the graph $K_{3,3}$ shown in Fig. 1.1b: The graph admits a 2-coloring that assigns one color to the vertices $U = \{u_1, u_2, u_3\}$ and another color to the vertices $W = \{w_1, w_2, w_3\}$. The 2-coloring decomposes the vertices $K_{3,3}$ into two independent sets, U and W, and induces a single Kempe-component. Given a coloring, a *Kempe-exchange* operation swaps the color assignments within a Kempe-component.

Matchings A *matching* in a graph G is a subset $M \subseteq E(G)$ such that each node of G is incident to at most edge in M. A matching is called *perfect*, if each vertex of G is incident to exactly one edge in the matching. The thick edges in Fig. 1.1b are a perfect matching in the graph $K_{3,3}$. The problem of finding a matching of maximum cardinality in a graph can be solved in polynomial time, see e.g. [PS98, Chapter 10].

Chapter 2
The University Course Timetabling Problem

In this chapter, we study the University Course Timetabling Problem (UCTP), which is a very basic model of the course timetabling task. Despite its simplicity, the model captures the essence of many more complex models: for example, it serves as foundation of the course timetabling models studied in Chaps. 3 and 4. In the following, we will consider two problems based on the UCTP model, a search problem and a reconfiguration problem. The model considers as entities events, rooms, timeslots and event conflicts. A timetable assigns to each event a room and a timeslot. Additionally, a timetable should have two basic properties: First, no two lectures are assigned to the same room and the same timeslot, and second, no two lectures marked as conflicting are scheduled simultaneously, i.e., in the same timeslot. Given appropriate data, the search problem asks for a *witness timetable* that satisfies all required properties. Such a timetable is called feasible. Given two feasible timetables, the reconfiguration problem asks if one timetable can be transformed into the other timetable step-by-step using some elementary operation, such that all intermediate timetables are also feasible. We present a theoretical investigation of the reconfiguration problem and an experimental evaluation of a specialized heuristic for the search problem.

2.1 Problem Formulation

The University Course Timetabling Problem (UCTP) formalizes the task of creating a weekly course schedule for students and lecturers at a university. The UCTP is equivalent to the one given in [dW85, Section 3.4]. Our purpose here is to establish the general setting and to introduce the terminology. A discussion of related problems such as the school timetabling problem and the examination timetabling problem is postponed to the next section.

M. Mühlenthaler, *Fairness in Academic Course Timetabling*, Lecture Notes in Economics and Mathematical Systems 678, DOI 10.1007/978-3-319-12799-6_2

The UCTP is the following search problem:

Definition 2.1 (University Course Timetabling Problem (UCTP))
INSTANCE:

- a set of events $E = \{e_1, \ldots, e_n\}$
- a set of rooms $R = \{r_1, \ldots, r_\ell\}$
- a set of timeslots $P = \{p_1, \ldots, p_k\}$
- a graph $G = (E, L)$ with nodes E and edges $L \subseteq \{\{u, v\} \mid u, v \in E\}$

The graph G is referred to as the *conflict graph*. Two events are called *conflicting* if they are adjacent in G. The set $P \times R$ contains the *resources*. A *timetable* τ is an assignment $\tau : E \to P \times R$. Two events e, e' are *overlapping*, if $e \neq e'$ and $\tau(e) = \tau(e')$. A timetable is called *overlap-free* if no two events overlap. Two events e, e' are *clashing* in τ, if they are conflicting and they are assigned to the same timeslot. A timetable is *feasible*, if it is clash-free and overlap-free.
TASK: Find a feasible timetable.

Deciding if there is a clash-free timetable is NP-complete for $|P| \geq 3$. This can be shown by a straight-forward reduction from the (vertex-)k-coloring problem.

The clash-freeness requirement and its relation to the vertex coloring problem is the heart of the matter of automated timetabling in the academic context. It occurs in various formalizations of course and exam timetabling tasks, see e.g. [Sch99]. Depending on the context and the regulatory framework, numerous other requirements may arise in practice, mostly related to resources being available only under certain conditions. We will discuss two types of such requirements namely *availability requirements* and *precedence requirements*, which can be added to the basic UCTP formulation above. Such requirements often arise in practice, see e.g. [Car01, SH07]. Availability requirements dictate that only a certain subset of the resources can be assigned to an event. The availability function α determines for each event the set of available resources:

$$\alpha : E \to \mathcal{P}(P \times R) \ .$$

A timetable τ satisfies the availability requirements, if for each event $e \in E$, $\tau(e) \in \alpha(e)$. For convenience, the availability function is typically specified by marking rooms and timeslots as (un)available individually per event, see generally [Sch99]. For example, if a timeslot p is marked as unavailable for an event e, then $\alpha(e)$ contains no resources that refer to p. A timetable τ satisfies the timeslot availability requirements, if for each event $e \in E$, $\tau(e)$ does not refer to a timeslot that has been marked as unavailable for e. Similarly, if a room r is unavailable for e, then $\alpha(e)$ does not contain any resources that refer to r. A timetable τ satisfies all room availability requirements, if for each event $e \in E$, $\tau(e)$ does not refer to a room that has been marked as unavailable for e. Clearly, a timetable satisfying the room and timeslot availability requirements satisfies the availability requirements.

Precedence requirements mandates that certain events need to be take place before other events in the weekly schedule. Let the timeslots $(P, <)$ be ordered and the resources $(P \times R, <)$ be ordered according to the timeslots, i.e., for two resource $(p,r),(p',r') \in P \times R$, $(p,r) < (p',r')$ iff $p < p'$. The precedence relation $\prec \subseteq E \times E$ on the events specifies the precedence requirements: For two events $e, e' \in E$, if $e \prec e'$ then e should precede e' in the timetable according to the given ordering of P. A timetable τ satisfies the precedence requirements, if for all events $e, e' \in E$, $e \prec e' \Rightarrow \tau(e) < \tau(e')$.

An important subproblem of the UCTP is the *room assignment problem*. Given a timeslot $p \in P$, a set $E' \subseteq E$ of events admits a room assignment, if there is an injective mapping $\rho : E' \to R$ such that $(p, \rho(e))$ is available for each $e \in E'$. If the choice of rooms is not restricted by availability requirements, then any set $E' \subseteq E$ of cardinality at most $|R|$ admits a room assignment. On the other hand, if availability requirements restrict the room choices then E' admits a room assignment if the following bipartite graph has a matching of cardinality $|E'|$:

$$H = (E' \mathbin{\dot\cup} R, F) \ ,$$

where $F \subseteq \{\{e,r\} \mid e \in E, r \in R\}$ and $\{e,r\} \in F$ if and only if (p,r) is available for the event e.

In addition to the basic entities such as rooms, timeslots, and events, there are typically some compound structures that represent groups of entities, see e.g. [DGMS07]. These compound structures do not add substantial content to the UCTP. However, depending on the setting, they may permit a more intuitive problem description. This is particularly useful for the specification of the objective function in optimization variants of the UCTP.

We will now give examples of compound structures, some of which will occur in the problems studied in Chaps. 3 and 4. Periods are typically grouped into *days* and *timeslots* in order to aid the specification of pattern and precedence constraints, see e.g. [MSP+10]. Furthermore, the conflict graph may be specified in terms of *courses* together with either *curricula* or an *attendance matrix* [MSP+10]. A *course* is a nonempty set of events and the courses are a partition of the events. The individual events of a course are sometimes referred to as *lectures*. Ordinarily, any two events of a course are assumed to be conflicting. A curriculum is a set of courses whose events must not be scheduled simultaneously. Thus, a curriculum induces a clique in the conflict graph. If students are required to explicitly enroll in a set of courses, an attendance matrix can be used to specify which student is enrolled in which course and from this information the conflict graph can be deduced. Additional conflicts may be due to a set of courses being taught by the same lecturer.

2.2 Related Problems

In the literature, numerous combinatorial problems can be found which formalize the task of creating timetables in the educational context. The educational timetabling problems are generally divided into three broad categories, school timetabling, course timetabling, and examination timetabling [Sch99]. In addition to arranging courses and exams in a timetable, there is the task of providing students with an adequate and balanced workload during their course of study. This has been formalized as the Balanced Academic Curriculum Problem (BACP). Furthermore, there is the task of creating groups of students for tutorials, lab exercises, and the like. This problem often appears as a subproblem of the UCTP and is known as the Student Sectioning Problem (SSP). We will give a brief overview over these problems in the following sections.

We will focus on structural aspects of these problems and summarize the known complexity results. Our motivation for this is twofold: First, we intend to highlight relations between the various problems on a basic combinatorial level. Second, the analysis of the connectedness of the UCTP search space presented in Sect. 2.3.2 considers the reconfiguration variant of the graph coloring problem (see Sect. 2.3.1), which is closely related to the UCTP. The basic formulations of the problems mentioned above give some insights into which reconfiguration problems could be considered for a similar analysis of the search space for these problems. In fact, due to the similar structure, the search space analysis we present in Sect. 2.3.2 applies to the Examination Timetabling Problem (ETP) as well.

Most variants of the problems mentioned above are hard in the sense that the decision variants of the basic problem formulations are NP-complete. The two exceptions are the School Timetabling Problem (STP) and the SSP. Polynomial time algorithms are known for the decision and search variants of these two problems. However, by introducing additional constraints such as timeslot availability requirements, these problems also become NP-complete [dW85, DPS13]. The reductions given by Cooper and Kingston show that there are in fact several sources of NP-completeness for different flavors of the timetabling problem [CK95].

2.2.1 School Timetabling

The school timetabling problem captures the task of assigning teachers to classes in a school. Despite the similar educational context, the STP exhibits a combinatorial structure that is quite different from the UCTP. For the purpose of a brief discussion of the STP, we give a problem formulation that is equivalent to the one proposed in [dW85]. In this model we assume that for each class, the teachers as well as the required number of lectures taught by them are known in advance. For a comprehensive overview of different variants and extensions of the STP as well as solution approaches from the literature please refer to the survey articles [Sch99, Pil13].

Definition 2.2 (School Timetabling Problem (STP))
INSTANCE:

- a set of teachers $T = \{t_1, \ldots, t_n\}$
- a set of classes $C = \{c_1, \ldots, c_m\}$

- a set of timeslots $P = \{p_1, \ldots, p_k\}$
- a $m \times n$ requirement matrix $M = (m_{ij})$, where m_{ij} is the number of lectures taught by teacher t_j in class c_i

Let E be the set of lectures. The number of lectures is determined by M: there are r_{ij} distinct lectures taught by t_j for each class $c_1, \ldots, c_m$. A *meeting* is an element of $T \times C$. A timetable is an assignment $\tau : E \to P$. A timetable is *feasible* if each teacher and each class have at most one lecture in each timeslot.
TASK: Find a feasible timetable.

For a given STP instance $\mathcal{I}$, there is a feasible timetable if and only if each teacher and each class has at most k lectures [dW97]. This result is a consequence of Kőnig's theorem on the existence of k-edge colorings in bipartite multigraphs: every bipartite multigraph G has a k-edge coloring, where k is the maximum degree of the graph G [Kőn16]. Let G be the bipartite multigraph with nodes $T \cup C$ and r_{ij} multiedges between t_j and c_i for each $j \in \{1, \ldots, n\}$, $i \in \{1, \ldots, m\}$. By applying Kőnig's theorem the result is immediate. Equivalently, a feasible timetable can be constructed by finding a sequence of k edge-disjoint matchings in a k-regular bipartite multigraph that is obtained by adding appropriate dummy nodes and edges to G [EIS76]. Thus, in contrast to the UCTP, the search problem STP can be solved in polynomial time. However, if availability requirements are added to the model, which prohibit the use of certain timeslots for teachers, classes or individual meetings, then deciding the existence of a feasible timetable is NP-complete [EIS76, dW97].

In addition to availability requirements, examples of practical requirements that may be included as hard or soft constraints are preference, precedence, workload and lecture distribution requirements. For a systematic review of the different types of constraints that occur in the various STP formulations see e.g. [Pil13]. Due to the similar educational context there is a considerable congruence of UCTP and STP requirements. However, there are certainly differences in the administrative and legal requirements, which may impact, for example, the nature of workload and lecture distribution constraints. For example, leaving students unattended is more problematic in a school than in a university setting [Sch99].

2.2.2 Examination Timetabling

The Examination Timetabling Problem (ETP) and the UCTP are very similar on a structural level. In fact, the UCTP formulation given in Definition 2.1 can serve as a basic model of examination timetabling problems: Each event corresponds to an exam and the task is to schedule all exams such that clash and overlap freeness requirements are satisfied. According to Schaerf, "it is difficult to make a clear distinction between the two problems" [Sch99]. Assuming our UCTP model from

Sect. 2.1, the following three characteristics of the ETP from [Sch99] distinguish it from the UCTP:

- There are different kinds of constraints, in particular distribution constraints that limit the number of exams per day and require that exams are spread over the available timeslots.
- The number of timeslots may be variable.
- Several exams may share a room.

For an overview of different approaches to solving the ETP found in the literature, please refer to the recent survey articles [Lew08, QBM+09].

2.2.3 *Other Related Problems*

In this section, we will discuss the SSP and the BACP, two combinatorial problems that, like the UCTP, appear in the academic setting. From the organizational perspective, the three problems are related in the following way: The academic workload determined by the BACP for a curriculum defines conflict constraints of the UCTP, and a feasible timetable for some UCTP instance serves as input of the student sectioning.

The Student Sectioning Problem

In order to deal with courses that are attended by a large number of students, universities may split up courses into several sections. Given a fixed timetable, the Student Sectioning Problem (SSP) captures the task of assigning students to sections of the courses such that the course structure is respected and conflicts are minimized [Sch99]. A task that often occurs in practice is the assignment of students to tutorial sessions with a limited capacity of each session. The SSP can be considered as a subtask in the course timetabling process. In recent course and examination timetabling models for benchmarking, such as the ones from the International Timetabling Competition 2007 (ITC2007) [MSP+10], student sectioning is not a part of the problem model. Rudová et al. argue however, that for large-scale practical course timetabling, it is necessary to consider sectioning as a part of the overall course timetabling problem [RMM11].

Recently, several complexity results have been established for the SSP by Dostert et al. [DPS13]: Consider the following Basic Student Sectioning Problem (BSSP): Let $C = \{c_1, \ldots, c_n\}$ be a set of courses, $P = \{p_1, \ldots, p_k\}$ be a set of timeslots, and let $B = (b_{ij})$, be the capacity matrix, where $i \in \{1, \ldots, n\}$, $j \in \{1, \ldots, k\}$. Each entry b_{ij} of the matrix gives the maximum capacity of the section of course c_j in timeslot i. Assuming that each student is enrolled in all n courses of a given timetable, the task is to decide if there is a sectioning for m students such that no student is assigned to two simultaneous sections and all capacities requirements are respected. The BSSP is equivalent to deciding the existence of m edge-disjoint matchings of cardinality n in a bipartite multigraph [DPS13]. The BSSP can be solved in polynomial time, but adding any combination of the following three kinds of requirements makes the problem NP-complete [DPS13]:

- Students can choose from several courses and their choices must be respected.
- Students have timeslot availability requirements.
- Sections have multiple events in different timeslots and conflicts between sections of different events must be respected.

The SSP and the STP have a similar combinatorial structure in the sense that both problems can be modeled as network flow problem with sources and sinks appropriately connected to the weighted complete bipartite graphs given by the capacity matrix and the requirements matrix, respectively. The main difference is the query: The SSP consists of deciding if the flow constraints are maintained for a particular number of students in the model. The STP asks if the arcs generated by the requirements matrix have left-over capacity for a given number of timeslots.

The Balanced Academic Curriculum Problem

Another problem related to course timetabling that occurs in the academic context is the BACP proposed in [CM01]. The general task is to assign courses to teaching periods (e. g. terms) such that the workload is evenly distributed over the teaching periods. In a sense, a solution to the BACP determines which courses need to be scheduled in a given semester, and thus a part of the input of the course timetabling task. An instance $\mathcal{I}$ of a very basic formulation consists of a set C of courses, a set P of teaching periods and a credit assignment $C \rightarrow \mathbb{N}$. By a reduction from the 3-partition problem [GJ79, Problem SP15], Chiarandini et al. showed that deciding whether $\mathcal{I}$ admits an assignment of courses to teaching periods such that the workload is perfectly balanced is strongly NP-complete [CDGGS12]. More realistic problem formulations include prerequisite requirements for courses, lower and upper limits for the workload (credits and courses) in each teaching period [CM01, HKW02].

We will revisit the BACP in Chap. 3 in the context of adding fairness criteria to optimization problems.

2.3 The Search Space

We provide a description of the structure of the search space of the UCTP with timeslot availability requirements in terms of graph theoretic concepts. Our motivation of studying the connectedness of the search space stems from the observation that a number of solutions approaches solve course timetabling optimization problems in two steps, see [Lew06, Section 2.3.2]: The first task is to solve the search problem and the second task is to find among the feasible timetables an optimal one. We provide a brief overview of related results from the literature on the vertex coloring problem and its reconfiguration variant (*recoloring*, for short). We present some new results about the connectedness of so-called reconfiguration graphs which model the search space of clash-free timetables. We derive improved sufficient conditions for the connectedness of the clash-free timetable and show that for a range of instances the clash-free timetables are indeed connected.

2.3.1 Vertex Coloring and Recoloring

For the UCTP and the ETP, the chromatic number of the conflict graph provides a lower bound on the number of timeslots that are required for a feasible schedule. For $k \geq 3$, the problem of deciding whether a graph has a k-coloring is NP-complete [GJ79]. Computing the chromatic number of a graph G with n nodes is NP-hard and for any $\epsilon > 0$, the chromatic number is NP-hard to approximate within $O(n^{1-\epsilon})$ [Zuc07]. However, there are (polynomial-time computable) upper bounds on the chromatic number that prove to be useful in the context of timetabling. The bounds result from analyzing the *greedy coloring algorithm*.

The Greedy Coloring Algorithm

Given a graph G and an ordering of its vertices, the greedy coloring algorithm sequentially colors the vertices in the given order, such that a vertex v is assigned to the smallest color that does not currently appear in the neighborhood of v. For any vertex ordering, the greedy coloring algorithm uses at most $\Delta(G) + 1$ colors, since each vertex is adjacent to at most $\Delta(G)$ neighbors (each of which may have a different color). Better bounds can be obtained by carefully choosing a vertex ordering. Welsh and Powell proposed to sort the vertices in non-increasing order according to their degrees [WP67]. Greedily coloring the vertices of G in this order yields a coloring with at most

$$\text{wp}(G) = \max_{i \in \{1,\ldots,n\}} \min\{d_i, i\} \tag{2.1}$$

colors, where d_i is the degree of the ith vertex. They used this result to derive an improved bound on the number of timeslots required for creating clash-free schedules in the context of examination timetabling.

A further improvement is due to the analysis of the greedy coloring algorithm by Szekeres and Wilf [SW68] and Matula [Mat68]. A graph G is called k-degenerate, if its vertices can be ordered such that each vertex has at most k neighbors preceding it. The smallest k for which G admits such an ordering is the *degeneracy* $\deg(G)$. Let $\sigma(G)$ be an ordering of the vertices of G that is a witness for the degeneracy of

G. The greedy coloring algorithm uses at most $\deg(G) + 1$ colors, if the vertices of G are processed sequentially according to $\sigma(G)$. The degeneracy is equivalent to the largest minimum degree of any subgraph, that is,

$$\deg(G) = \max_{H \subseteq G} \delta(H) \ . \tag{2.2}$$

A witness vertex ordering σ for $\deg(G)$ can be constructed efficiently by repeatedly removing vertices of minimal degree and then reversing the order of removal [Mat68, SW68]. By Using appropriate data structures, $\deg(G)$ can be computed along with a witness σ in time $O(\max\{|V(G)|, |E(G)|\})$ [BZ03].

Proposition 2.3 *The following inequalities hold for any graph G:*

$$\chi(G) \leq \deg(G) + 1 \leq \mathrm{wp}(G) + 1 \leq \Delta(G) + 1 \ .$$

Proof 1. $\chi(G) \leq \deg(G) + 1$. We essentially replicate the proof from [SW68]. Let $G' \subseteq G$ such that $\chi(G') = \chi(G)$, but for any subgraph $H \subset G'$ we have $\chi(H) < \chi(G)$. Thus, G' cannot contain a vertex with fewer than $\chi(G') - 1$ neighbors. Due to Eq. (2.2),

$$\chi(G') - 1 \leq \delta(G') \leq \deg(G) \ .$$

2. $\deg(G) \leq \mathrm{wp}(G)$. From Eq. (2.1) we can conclude that any graph G is $\mathrm{wp}(G)$-degenerate. Since $\deg(G)$ is smallest number k such that G is k-degenerate, we have $\mathrm{wp}(G) \leq \deg(G)$.
3. $\mathrm{wp}(G) \leq \Delta(G)$. The bound $\Delta(G) + 1$ on the number of colors used by the greedy coloring algorithm holds for any ordering of the vertices, including the ones that list the vertices of G in non-increasing order according to their degrees. Therefore, $\mathrm{wp}(G) \leq \Delta(G)$.

□

In Sect. 2.4 we will discuss heuristic approaches to solving the UCTP that are based on the greedy coloring algorithm. Some of these heuristics determine the vertex ordering on-the-fly, which may lead to an improvement in practice but does not to lead to improved general bounds on the chromatic number. For an overview of sequential graph coloring heuristics with applications to timetabling, see [BdWK03].

Vertex Coloring Reconfiguration Graphs

In order to gain some insight into the structure of the search space of the UCTP, we consider the following problem: Given a graph G, is it possible to transform any coloring of G into any other coloring of G in a step-by-step manner, such that each intermediate coloring is proper. Problems of this kind are closely related to reconfiguration problems. Given an instance $\mathcal{I}$ of a reconfiguration problem, the adjacency relation $\sim$ on $\mathcal{S}(\mathcal{I})$ gives rise to a *reconfiguration graph*, whose nodes are the feasible solutions $\mathcal{S}(\mathcal{I})$ and two nodes $u, v \in \mathcal{S}(\mathcal{I})$ are adjacent iff $u{\sim}v$. In graph-theoretic terms, a reconfiguration problem addresses the question whether two nodes of the reconfiguration graph are in the same connected component.

There has recently been quite some interest in the study of reconfiguration variants of classical combinatorial problems and their corresponding reconfiguration graphs. A central theme in this line of research is to determine the relation between the complexity class of a combinatorial problem and the complexity class of its reconfiguration problem. Many reconfiguration problems of NP-complete problems turn out to be PSPACE-complete, including for instance Boolean satisfiability (SAT) [GKMP09], the independent set, clique, vertex cover, set cover and integer programming reconfiguration problems [BC09, IDH$^+$11], as well as the graph k-coloring reconfiguration problem for $k \geq 4$ [BC09]. An exception is the reconfiguration variant of the 3-coloring problem, which is in P [CvdHJ11]. Most reconfiguration variants of problems in P studied in the literature have been shown to be in P as well. For instance, the reconfiguration variants of polynomial-time solvable SAT problems [GKMP09] and the matching problem [IDH$^+$11]. An exception is the shortest path reconfiguration problem, which is PSPACE-complete [Bon12].

In addition to attacking reconfiguration problems from the complexity theoretic point of view, there have been investigations of structural properties of reconfiguration graphs [Moh07, BJL$^+$11, BJL$^+$12]. The analysis of the UCTP search space will follow this line of research. Given a coloring c of a graph G, an *elementary recoloring* changes the color of a single vertex of G. Two k-colorings c_1 and c_2 of G are adjacent, $c_1{\sim_E}c_2$, if there is an elementary recoloring that transforms c_1 into c_2. The Kempe-exchange is a generalization of the elementary recoloring operation. Recall that, given two colors a and b, a Kempe-exchange switches the colors of a Kempe-component, which is a connected component in $G(a, b)$. The result of this operation is a new coloring, such that, within the Kempe-component, all vertices of the of color a are assigned to color b and vice versa. Two colorings c_1 and c_2 of G are adjacent with respect to the Kempe-exchange, $c_1{\sim_K}c_2$, if there is a Kempe-exchange that transforms c_1 into c_2. Each of the two adjacency relations $\sim_E$ and $\sim_K$ gives rise to a graph structure on the set of k-colorings of G.

Definition 2.4 ((Kempe-)k-Coloring Graph) For a graph $G = (V, E)$ and $k \in \mathbb{N}$ let

$$\mathcal{V} := \{c : V \to \{1, \dots, k\} \mid c \text{ is a } k\text{-coloring of } G\}$$

$$\mathcal{E}_E := \{\{c_1, c_2\} \mid c_1, c_2 \in \mathcal{V} \text{ and } c_1{\sim_E}c_2\}$$

$$\mathcal{E}_K := \{\{c_1, c_2\} \mid c_1, c_2 \in \mathcal{V} \text{ and } c_1{\sim_K}c_2\} \ .$$

Then the k-coloring graph is the graph $\mathcal{C}_k(G) = (\mathcal{V}, \mathcal{E}_E)$. The Kempe-$k$-coloring graph is the graph $\mathcal{K}_k(G) = (\mathcal{V}, \mathcal{E}_K)$.

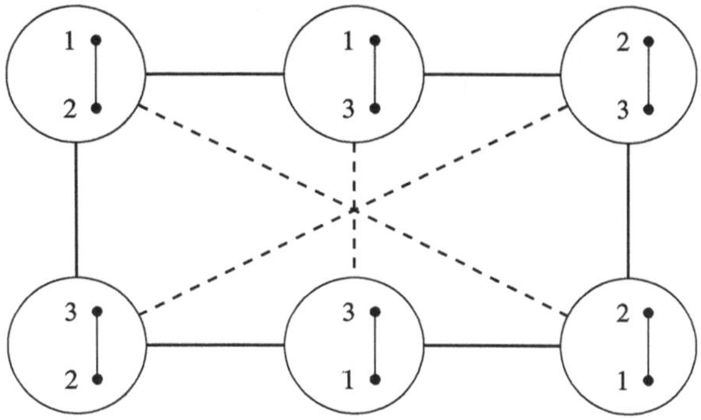

Fig. 2.1 The (Kempe-)3-coloring graph of K_2. *Solid edges* correspond to elementary recolorings. *Dashed edges* correspond to Kempe-exchanges that are not equivalent to an elementary recoloring

Since the Kempe-exchange is a generalization of the elementary recoloring operation, $\mathcal{C}_k(G) \subseteq \mathcal{K}_k(G)$ holds for any graph G and any $k \in \mathbb{N}$. Figure 2.1 shows the reconfiguration graphs $\mathcal{K}_3(K_2)$ and $\mathcal{C}_3(K_2)$. Any two nodes that are connected by a (thick) solid line are adjacent in $\mathcal{K}_3(K_2)$ and $\mathcal{C}_3(K_2)$. Any two nodes that are connected by a dashed line are adjacent in $\mathcal{K}_3(K_2)$ but not in $\mathcal{C}_3(K_2)$. To the best of our knowledge, the recognition problem for reconfiguration graphs in general and for vertex coloring reconfiguration graphs in particular has not been addressed so far in the literature. We state the following observation:

Proposition 2.5 *Let K_n be the clique on n nodes. Then*

1. *$\mathcal{C}_k(K_n)$ is $(n(k-n))$-regular for all $k \geq n$,*
2. *$\mathcal{C}_{n+1}(K_n)$ has girth six for all $n \geq 2$,*
3. *$\mathcal{C}_k(G)$ has girth three for any non-empty graph G and $k > \delta(G) + 2$.*

Proof Part 1: For each of the n nodes of K_n there are $k-n$ colors to choose from while all other colors remain fixed, so each node has $n(k-n)$ neighbors.
Part 2: There is no cycle in $\mathcal{C}_k(K_n)$ that involves a color change of only a single vertex since $k = n + 1$. Any cycle in $(\mathcal{C}_{n+1}(K_n))$ that changes the colors of at least three vertices of K_n has length at least six, since each vertex of G has to be assigned its original color eventually. Consider cycles of $\mathcal{C}_{n+1}(K_n)$ that recolor exactly two vertices. Such cycles are isomorphic to $\mathcal{C}_3(K_2)$, which is a cycle of length six, see Fig. 2.1. Thus, $\mathcal{C}_{n+1}(K_n)$ contains cycles of length six, and all other cycles have at least length six. Therefore, $\text{girth}(\mathcal{C}_{n+1}(K_n)) = 6$.
Part 3: Pick any coloring c of G and a vertex $u \in V(G)$ of degree $\delta(G)$. Since $k > \delta(G) + 1$ there are at least two colors α and β that are not present in the neighborhood of u. Therefore, u can be recolored to α, then to β and back to $c(u)$, which implies that $\text{girth}(\mathcal{C}_k(G)) = 3$. □

An inspection of the spectrum of $\mathcal{C}_{k+1}(K_k)$ for small values of k leads us to the following conjecture:

Conjecture 2.6 For $k \geq 3$, the k-coloring graph $\mathcal{C}_{k+1}(K_k)$ has eigenvalue spectrum $0, \pm 1, \dots, \pm k$.

The Kempe-exchange is a popular operation for exploring the search space of course and examination timetabling problems. For this purpose, we are in particular interested in whether a (Kempe-)k-coloring graph is connected, and to some extent, in upper bounds on the number of recoloring operations needed to reach any coloring from a given starting point. Clearly, if $\mathcal{C}_k(G)$ is connected for some k and G, then $\mathcal{K}_k(G)$ is also connected. Bonsma and Cereceda showed in [BC09] that deciding if two colorings c, c' of a graph G are connected in $\mathcal{C}_k(G)$ is PSPACE-complete for $k \geq 4$ by giving a reduction from the SLIDINGTOKENS problem from [HD05]. However, the connectedness of $\mathcal{C}_k(G)$ has been established for sufficiently large k:

Theorem 2.7 *The k-coloring graph $\mathcal{C}_k(G)$ is connected, if*

1. *$k \geq \Delta(G) + 2$ ([Jer95]), or*
2. *$k \geq \deg(G) + 2$ ([BC09]).*

Bonsma and Cereceda showed in [BC09] that two connected k-colorings may have superpolynomial (in the size of the graph) distance in the k-coloring graph. In the following cases however, the diameter of the reconfiguration graph is known to be quadratic:

Theorem 2.8 *The k-coloring graph $\mathcal{C}_k(G)$ has diameter $O(|V(G)|^2)$ if any of the following conditions is satisfied:*

1. *$k \geq 2\deg(G) + 1$ [BC09]*
2. *$\chi(G) = 3$ and $\mathcal{C}_3(G)$ is connected [CvdHJ11].*

Concerning Kempe-k-coloring graphs, Mohar has shown that:

Theorem 2.9 ([Moh07, Corollary 4.5]) *The Kempe-k-coloring graph $\mathcal{K}_k(G)$ is connected if G is planar and $k \geq \chi(G)$.*

The following theorem by Las Vergnas and Meyniel establishes sufficient conditions for the connectedness of Kempe-k-coloring graphs in a more general setting. We will give a proof which is essentially the same as the one given in [Moh07, Proposition 2.4]. However, we will present it in a slightly more algorithmic style. We encode a Kempe-exchange operation by a triple (a, b, u), where a and b are colors and u is a vertex of the Kempe-component. In order to conveniently modify colorings in our algorithms, we encode them as arrays indexed by the nodes of a graph. For example, by $c[u]$ we refer to the color of vertex u with respect to the coloring c.

Theorem 2.10 ([LVM81, Proposition 2.1]) *For any graph G, the Kempe-k-coloring graph $\mathcal{K}_k(G)$ is connected if $k \geq \deg(G) + 1$.*

Proof Let $\ell = \deg(G)$ and let c_1, c_2 be k-colorings of G. Without loss of generality, we assume that c_1 is the source coloring and c_2 is the target coloring. From a vertex ordering that is a witness of $\deg(G)$ we get a labeling $v_1, \dots, v_n$ of the vertices of G such that for each $i \in \{1, \dots, n\}$, v_i has at most ℓ neighbors in $G[v_1, \dots, v_i]$. We prove that the algorithm KEMPERECOLOR with input G, c_1, c_2, and the labeling

Algorithm 1: KEMPERECOLOR

input : graph G, labeling $v_1, \dots, v_n$ of the vertices, k-colorings c_1, c_2 of G
output: list of Kempe-exchanges transforming c_1 into c_2
data : array c of length n storing the current color of each vertex, list K of Kempe-exchanges

`/* initialization */`
$K \longleftarrow$ empty list
for $i \longleftarrow 1$ **to** n**:**
  $c[i] \longleftarrow c_1(v_i)$

`/* recoloring */`
for $i \longleftarrow 1$ **to** n**:**
  $G' \longleftarrow G[v_1, \dots, v_i]$
  `/* Kempe-exchange` κ `encoded as` $\kappa = (a, b, u)$`, where` a,b `are colors and` $u \in V(G')$ `*/`
  for $\kappa = (a, b, u) \in K$**:**
    let w.l.o.g. $c[i] \neq a$
1   **if** $c[i] = b$ *and* v_i *has exactly one neighbor of color* a *in* G'**:**
      $c[i] \longleftarrow a$
2   **else if** $c[i] = b$ *and* v_i *has at least two neighbors of color* a *in* G'**:**
      choose color $b' \neq b$, which is not used by any neighbor of v_i in G'
      insert Kempe-exchange (b, b', v_i) right before κ in K
      $c[i] \longleftarrow b'$
3   append Kempe-exchange $(c_2(v_i), c[i], v_i)$ to K
return K

$v_1, \dots, v_n$ outputs a sequence of Kempe-exchanges that transforms c_1 into c_2. For this purpose, we show that for each iteration i of KEMPERECOLOR the following two properties hold: (i) the colors of $v_1, \dots, v_{i-1}$ are not altered due to v_i, and (ii) v_i is assigned to its target color $c_2(v_i)$ at the end of iteration i.

The crucial steps needed to establish property (i) are performed in lines 1 and 2. If v_i is not affected by Kempe-exchange $\kappa = (a, b, u)$ then no special treatment of v_i is needed. If v_i is affected by κ, then there are two cases to consider. If v_i has color b and there is exactly one neighbor of v_i in $G' = G[v_1, \dots, v_i]$ of color a (line 1), then κ either changes the color of v_i to a or it leaves the color of v_i unchanged. On the other hand, if v_i has color b and at least two neighbors of v_i in G' have color a (line 2), then at most $\ell - 1$ colors are present in the neighborhood of v_i in G'. Therefore, if $\ell + 1$ colors are available, then v_i can be recolored to a color b' that is not present in the neighborhood of v_i prior to performing the Kempe-exchange κ. As a consequence, v_i is then unaffected by κ and thus, v_i cannot interfere with the colors of $v_1, \dots, v_{i-1}$. At the end of iteration i (line 3), v_i can safely be assigned to color $c_2(v_i)$, which establishes property (ii).

Thus, after iteration n, each vertex has been assigned to its target color. At most $\ell + 1$ colors are required in each step. □

Note that the number of Kempe-exchanges added to the list K in Algorithm 1 may double in each iteration in the worst case. Thus, although no witness of two colorings at a superpolynomial has been mentioned in the literature, we cannot conclude that the output of KEMPERECOLOR has polynomial size.

In addition to these results several properties of (Kempe-)k-coloring graphs are known for graphs belonging to particular classes. A graph is *chordal*, if any cycle of length at least four has a chord. Every chordal graph is $(\chi - 1)$-degenerate. A *perfect elimination ordering* of a graph is an ordering of the vertices such that for each vertex v the neighbors that occur after v in the ordering induce a clique. A graph is chordal if and only if it admits a perfect elimination ordering. A perfect elimination ordering of the vertices of a chordal graph can be found efficiently by a variant of Lexicographic Breadth-first Search (LBFS) [RTL76]. LBFS is a particular class of Breadth-first Search (BFS) algorithms that is useful for efficiently recognizing various graph classes, including for instance chordal graphs, chordal-bipartite graphs, interval graphs and co-graphs [Cor05, Ueh02]. A graph is *chordal bipartite*, if it is bipartite and any cycle of length at least six has a chord.

Theorem 2.11 ([BJL⁺11]) *If G is chordal or chordal-bipartite and $k \geq \chi(G) + 1$ then* $\mathrm{diam}(\mathcal{C}_k(G)) = O(|V(G)|^2)$.

Bonamy et al. further show that for each $k \geq 2$ there is a k-colorable chordal graph such that $\mathrm{diam}(\mathcal{C}_{k+1}(G)) = \Theta(|V(G)|^2)$ [BJL⁺11]. We will show that any chordal graph G can be recolored with at most $|V(G)|$ Kempe-exchanges, i.e., $\mathrm{diam}(\mathcal{K}_k(G)) \leq |V(G)|$ (see Corollary 2.13). For this purpose consider the sequential recoloring algorithm SEQRECOLOR shown in Algorithm 2: The input is a graph G, labeling of the vertices $V(G)$ and two colorings c, c' of G. We assume that c is the source coloring and c' is the target coloring. The algorithm recolors the vertices one-by-one according to the labeling such that in each step i, one Kempe-exchange is performed that assigns to the current vertex v_i its target color $c'(v_i)$ if its current color and its target color differ. The output is a coloring of G. If the output is equivalent to the target coloring c' we say that SEQRECOLOR transforms c into c'. We investigate the conditions under which this is the case.

Algorithm 2: SEQRECOLOR

```
input : graph G, colorings c, c' of G, labeling v_1, ..., v_|V(G)| of V(G)
output: the coloring b of G

/* initialization                                              */
for i <— 1 to |V(G)|: b[v_i] <— c(v_i)

/* recoloring                                                  */
for i <— 1 to |V(G)|:
    if b[v_i] ≠ c'(v_i):
        perform Kempe-exchange (b(v_i), c'(v_i), v_i)

return b
```

Theorem 2.12 *For any chordal graph G there is an ordering of $V(G)$ such that for any two colorings c, c' of G* SEQRECOLOR *transforms c into c'.*

Proof Without loss of generality, we can assume that G is connected, since if it is not, we can treat each connected component individually. Since G is chordal it admits a perfect elimination ordering. We show that applying SEQRECOLOR to the vertices of G in a reversed perfect elimination ordering transforms c into c'. We verify that in each iteration i, $1 \leq i \leq n$, the Kempe-exchange that recolors v_i from its current color to its target color $c'(v_i)$ leaves the color of any vertex $w > v_i$ unchanged.

We assume for a contradiction that in iteration i there is a vertex $w > v_i$ in the elimination ordering such that the Kempe-exchange $\kappa = (b[v_i], c'(v_i), v_i)$, which assigns v_i to its target color, changes the color of w as well. We pick w to be the smallest such vertex. If $b[v_i] = c'(v_i)$ then no recoloring is performed and thus no such w exists. If $b[v_i] \neq c'(v_i)$ then the Kempe-component corresponding to κ is a tree T, since $G(b[v_i], c'(v_i))$ is bipartite and every cycle of length at least four has a chord. Since the color of w is changed by κ it is a vertex of T and has either color $b[v_i]$ or $c'(v_i)$. Thus, there is a unique path $v_i = u_1 — \ldots — u_m — w$ in T. If w were adjacent to v_i then its color can neither be $b[v_i]$ nor $c'(v_i)$, so w is not adjacent to v_i and $m \geq 2$. The vertices on the path satisfy $u_m < \ldots < u_1 < w$ with respect to the elimination ordering, since $v_i = u_1 > u_2$ and for all $1 < i < m$, $u_i < u_{i-1} < u_{i+1}$ or $u_i < u_{i+1} < u_{i-1}$ implies that u_{i-1} and u_{i+1} are adjacent. From $u_m < u_{m-1} < w$ we can conclude that u_{m-1} is adjacent to w, which is a contradiction to T being a tree and thus, there is no vertex $w > v_i$ that is recolored by κ. □

The next corollary follows directly from Theorem 2.12 and the fact that SEQRECOLOR performs at most one Kempe-exchange per vertex.

Corollary 2.13 *If G is chordal and $k \geq \chi(G)$ then* $\operatorname{diam}(\mathcal{K}_k(G)) \leq |V(G)|$. □

Please note that the order in which the vertices are processed depends only on the structure of G, not on the given colorings. As a slight generalization of this, we introduce the class of sequentially recolorable graphs.

Definition 2.14 A graph G is *sequentially recolorable*, if for any two colorings c, c' of G there is an labeling of the vertices such that SEQRECOLOR transforms c into c'.

Clearly, due to Theorem 2.12 any chordal graph is sequentially recolorable. So far, no other graph classes are known to be sequentially recolorable. We conjecture that:

Conjecture 2.15 Every chordal bipartite graph is sequentially recolorable.

In the following, we will present some results, mainly about which methods cannot be used to prove Conjecture 2.15. The following proposition is a direct consequence of [Moh07, Proposition 2.1], which establishes that $\mathcal{K}_k(G)$ is connected for any $k \geq 2$ if G is bipartite. The proof is omitted in the reference, so we give a short proof here for completeness.

Proposition 2.16 *If G is bipartite and $k \geq 2$ then* $\text{diam}(\mathcal{C}_k(G)) \leq 2|V(G)|$.

Proof Let $G = (A \cup B, E)$ be a bipartite graph, c, c' be two k-colorings of G, $k \geq 2$, and b be a 2-coloring of G. Without loss of generality, we assume that G is connected and $b(v) = 1$ if $v \in A$, and $b(v) = 2$ otherwise. We further assume that the vertices of G are ordered such that $v > w$ whenever $v \in A$ and $w \in B$ and ties are broken arbitrarily. We show that SEQRECOLOR transforms c into b if the vertices are processed according to such an ordering. For this purpose, we show that whenever a Kempe-exchange is performed on a vertex $v \in A$ then the number of vertices of color 1 in A strictly increases. Assume for a contradiction that this is not the case. Then performing the Kempe-exchange that recolors v from its current color $\alpha \neq 1$ to 1 must also recolor another vertex $w \in A$ from color 1 to α. Thus, there is a path $v — \ldots — w$ of even length in the subgraph induced by the vertices of color 1 or α, and since G is bipartite vertices of color α and 1 must be alternating on the path. However, since v is of color α and w is of color 1 this is a contradiction. The same argument holds whenever a vertex in B is assigned the color 2. Thus, SEQRECOLOR transforms c into b.

Reversing the order of the Kempe-exchanges performed by SEQRECOLOR for transforming c' into b yields a sequence of Kempe-exchanges that transforms b into c'. Thus we can transform c into c' via the 2-coloring b. Since SEQRECOLOR is used at most twice, $\text{diam}(\mathcal{K}_k(G)) \leq 2|V(G)|$. □

It is an open question if the bound on the diameter of the reconfiguration graph can be improved to $\text{diam}(\mathcal{K}_k(G)) \leq |V(G)|$ if G is bipartite. Even for the class of chordal bipartite graphs the question is not settled (see Conjecture 2.15). In the spirit of the characterization of chordal graphs via a perfect elimination ordering, chordal bipartite graphs are characterized by the existence of a weak elimination ordering. A vertex u of a bipartite graph is called *weakly simplicial* if its neighbors can be labeled $v_1, \ldots, v_k$ such that $\text{N}(v_i) \subseteq \text{N}(v_j)$ whenever $i < j$ [BJL+12]. An ordering $v_1, \ldots, v_n$ of the vertices of a bipartite graph G is a *weak elimination ordering* if for each $i \in \{1, \ldots, n\}$, v_i is weakly simplicial in the subgraph of G induced by $\{v_i, \ldots, v_n\}$.

Theorem 2.17 ([Ueh02, Theorem 1]) *A bipartite graph is chordal bipartite if and only if its vertices admit a Weak Elimination Ordering (WEO).* □

A WEO of the vertices can be found in linear time by a variant of the LBFS algorithm proposed in [Ueh02]. Starting from a given initial vertex, generic BFS algorithm partitions the vertices of a graph into layers and processes the vertices layer by layer. Any two nodes in a layer have the same distance to the initial vertex. In contrast to the generic BFS, the LBFS algorithm visits the vertices of a layer in a specific order: In each step, the next vertex is selected from the vertices with the lexicographically largest label and the labels are updated after each iteration. The exact procedure is not relevant for our investigations. The details can be found

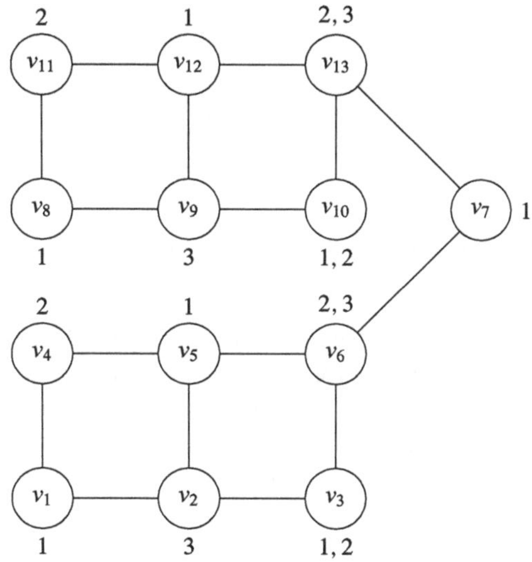

Fig. 2.2 A graph recoloring instance, consisting of a chordal bipartite graph G and two vertex colorings $c, c' : V(G) \to \{1, 2, 3\}$. The instance satisfies both parts of Proposition 2.18

e.g. in [Cor05]. However, reversing the ordering in which the vertices are visited by the LBFS yields a WEO.

We present a graph coloring reconfiguration instance such that for none of the vertex orderings produced by a BFS, SEQRECOLOR transforms the source coloring into the target coloring. In particular, this includes all WEOs produced by a LBFS traversal of the graph. However, we also give a WEO of the vertices for which SEQRECOLOR correctly transforms the source into the target coloring. Therefore, our instance does not rule out WEOs for proving that chordal bipartite graphs are sequentially recolorable.

Figure 2.2 shows a bipartite chordal graph G consisting of two dominoes connected via the node v_7. There are two colorings $c, c' : V(G) \to \{1, 2, 3\}$ shown next to the nodes. The first color is the one assigned by c and the second color is the one assigned by c'. If only one color $\alpha \in \{1, 2, 3\}$ is shown next to a node v then $c(v) = c'(v) = \alpha$. For example, $c(v_1) = c'(v_1) = 1$, $c(v_3) = 1$, and $c'(v_3) = 2$.

Proposition 2.18 *There is a chordal bipartite graph G and two colorings c, c' of G such that*

1. SEQRECOLOR *does not transform c to c' for any vertex ordering that is the revers of a BFS vertex ordering, and*
2. *there is a WEO for which* SEQRECOLOR *transforms c to c'.*

Proof First, consider a BFS traversal of the lower domino $G[v_1, \ldots, v_6]$, which begins at vertex v_6. The BFS partitions the vertices into the layers

$$\{v_6\}, \{v_3, v_5\}, \{v_2, v_4\}, \{v_1\} \ .$$

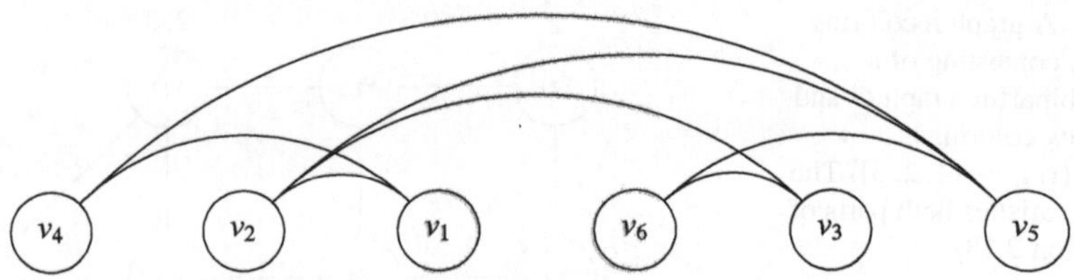

Fig. 2.3 A WEO of lower domino of the graph shown in Fig. 2.2

Breaking ties between the nodes within a layer is of no significance, thus the argument holds for any BFS specialization, including LBFS. After reversing the order of traversal, v_3 is processed by SEQRECOLOR after v_1, and therefore a Kempe-exchange assigning to v_3 its target color 2 also changes the color of v_5, which is in the previous layer. Thus, SEQRECOLOR will not produce the target coloring c'. By connecting the two dominoes via v_7, a BFS starting at any node either visits v_6 before v_3, or, equivalently, v_{13} before v_{10}. Therefore, SEQRECOLOR does not transform c into c' for any vertex ordering that is the reverse of a BFS vertex ordering, which concludes the proof of part 1.

Now, consider the vertex ordering $v_4, v_2, v_1, v_6, v_3, v_5$, shown in Fig. 2.3, which is a WEO of the lower domino. Similarly, $v_{11}, v_9, v_8, v_{13}, v_{10}, v_{12}$ is a WEO of the upper domino $G[v_8, \ldots, v_{13}]$. We combine the two vertex orderings such that the lower domino appears after the upper domino and insert v_7 in the ordering right after v_6. The resulting vertex ordering is a WEO. SEQRECOLOR transforms c into c', which proves part 2. □

2.3.2 *Connectedness*

According to the classification of heuristic optimization algorithms for timetabling problems in [Lew06], many approaches in the literature fall in the category of *two-step optimization algorithms*. In the first step, the underlying search problem is solved and in the second step, the optimization takes place, during which only feasible solutions are considered and the result of the first step is used as a starting point. A recent example of a state-of-the-art two-step approach is [LH10], numerous other examples can be found in [Lew06]. During the optimization step, feasible timetables are modified using Kempe-exchanges or similar operations that preserve the feasibility. It is natural to ask whether any feasible timetable, in particular an optimal one, can be reached from an initial feasible timetable. We investigate conditions that establish the connectedness of feasible timetables with respect to the Kempe-exchange.

In our analysis, we consider the clash-freeness requirement and do not include any requirements related to assigning rooms. We apply the results from Sect. 2.3.1 to show that for a range of benchmark instances (which include computer-generated and real-world instances) the search space of clash-free timetables is connected in the sense that the corresponding Kempe-p-coloring graph (see Definition 2.4) is connected. For other benchmark instances, the conditions for the connectedness are not satisfied, and further investigations are needed to settle whether the reconfiguration graphs are connected or not. We further show how timeslot availability requirements can be included in the analysis and derive improved conditions for the connectedness of the reconfiguration graphs in this setting.

In most applications, clash-freeness is not the only requirement a timetable needs to satisfy. Therefore, the overall goal is to include additional requirements such as timeslot availability, room availability and overlap-freeness in the analysis of the properties of the search space. Let $\mathcal{I}$ be a UCTP instance with p timeslots, a conflict graph G, and let V' be the set of colorings of G that satisfy the additional requirements. Then the search space of instance $\mathcal{I}$ is connected if $\mathcal{K}_p(G)[V']$ is connected. In particular, for the additional requirements above, the corresponding reconfiguration graphs are the subgraphs of $\mathcal{K}_p(G)$ induced by the following sets of nodes:

1. timeslot availability requirements:

$$V_\pi = \{c \in V(\mathcal{K}_p(G)) \mid \forall v \in V(G) : c(v) \text{ is available for event } v\}$$

2. overlap freeness and room availability requirements:

$$V_\rho = \{c \in V(\mathcal{K}_p(G)) \mid \forall i \in P : \text{ color class } i \text{ admits a room assignment}\}$$

In our analysis we will focus on conditions that establish the connectedness of $\mathcal{K}_p(G)[V_\pi]$. Regarding overlap freeness and room availability requirements, to the best of our knowledge, the properties of the corresponding reconfiguration graphs have not been studied so far. The *bounded vertex k-coloring problem* with bound $b \in \mathbb{N}$ is the problem of coloring a graph with k colors such that each color is used at most b times. The bounded vertex coloring problem has been studied, for example, by Lucarelli [Luc09], and Baker and Coffmann [BC96] in the setting of unit-time task scheduling on multiple processors and by de Werra in the timetabling context [dW97]. If overlap freeness is required and no particular room availability requirements are present, then the graph $\mathcal{K}_P(G)[V_\rho]$ is the reconfiguration graph of a bounded vertex coloring instance. The reconfiguration variant of the bounded vertex coloring problem seems to be an interesting problem which deserves further investigation. The situation gets more involved if room availability requirements are present. Checking if the k events in a color class admit a room assignment is equivalent to checking if a certain bipartite graph admits a matching of cardinality k. At the moment, a description of the corresponding reconfiguration graph that allows us to reason about its connectedness is out of reach.

Connectedness in the Presence of Period Unavailabilities

In the following, let $G = (V, E)$ be the conflict graph of a UCTP instance. The problem of deciding if there is a clash-free timetable that additionally satisfies timeslot availability requirements is equivalent to deciding if G admits a *list coloring*. Given a set $L(v)$ (called *list*) of available colors for each $v \in V$, a list coloring $c : V \to \bigcup_{v \in V} L(v)$ of G is a proper coloring of G such that $c(v) \in L(v)$ for each $v \in V$. Since in the UCTP context, the conflict graph is colored with the timeslots, the timeslot availability requirements for each event $v \in V$ can be expressed by letting $L(v)$ be the set of timeslots available for v. Graph coloring is a special case of list coloring, since all colors are available for each node. By using a standard technique [dW85, Proposition 3.2], list coloring can be reduced to graph coloring: Let the colors be labeled $1, \ldots, p$, where $p = \left|\bigcup_{v \in V} L(v)\right|$. Now, let the graph G' be a copy of G to which we add a clique C on p nodes $v_1, \ldots, v_p$. For each $v \in V(G)$, we add an edge $v — v_i$ to G', whenever $i \notin L(v)$. Clearly, G' admits a p-coloring if and only if G admits a list coloring.

Our goal is to show that $\mathcal{K}_p(G)[V_\pi]$ is connected if and only if $\mathcal{K}_p(G')$ is connected. However, there is no Kempe-exchange on $\mathcal{K}_p(G)[V_\pi]$ corresponding to a Kempe-exchange on G' involving any of the nodes $v_1, \ldots, v_p$. We construct a graph K, which is a copy of $\mathcal{K}_p(G)[V_\pi]$ with a self loop added to each node. Additionally, we add to K an edge between two colorings $u, v \in V(K)$, if there are two colors i and j such that u can be transformed into v by swapping the colors in all except a single connected component of $G(i, j)$. These additional edges are merely shortcuts for several individual Kempe-exchanges. Therefore, $\mathcal{K}_p(G)[V_\pi]$ is connected if and only if K is connected. Figure 2.4 shows the various graphs under consideration for

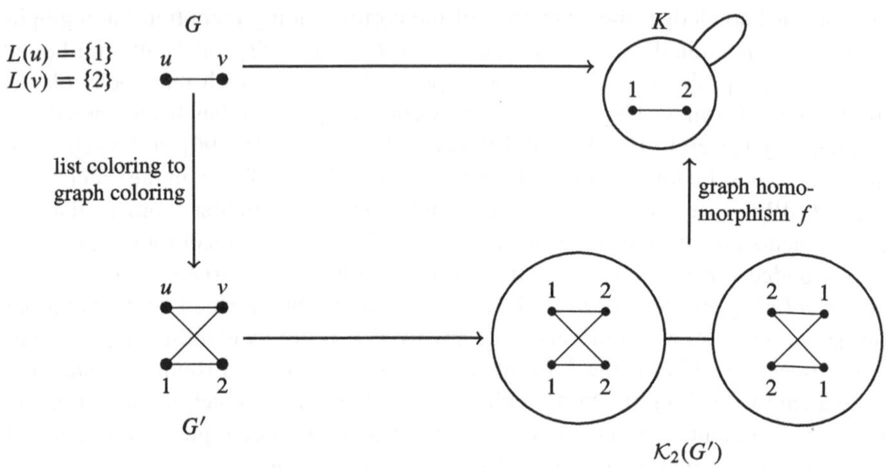

Fig. 2.4 Relations between the graphs G, G', K and $\mathcal{K}_p(G')$. The choice of G and the available colors determines the other graphs as described in the text. The existence of the graph homomorphism f is established by Lemma 2.19

a list-coloring instance consisting of a graph $G = (\{u, v\}, \{u — v\})$ and color lists $L(u) = \{1\}$ and $L(v) = \{2\}$. The nodes 1 and 2 of G' were added by the reduction from list to graph coloring.

Lemma 2.19 *There is a graph homomorphism* $f : \mathcal{K}_p(G') \to K$.

Proof We construct the mapping $f : V(\mathcal{K}_p(G')) \to V(K)$. Let $c \in V(\mathcal{K}_p(G'))$. First, we swap the colors of the p color classes such that v_i has color i for each $i \in \{1, \dots, p\}$. This can be achieved by applying a sequence of Kempe-exchanges to the coloring c: For each color $j \in \{1, \dots, p\}$, if the current color of v_j is $i \neq j$ we swap the colors in $G'(i, j)$. One Kempe-exchange is required for each connected component of $G'(i, j)$. Let c' be the resulting coloring. Except for the vertices $v_1, \dots, v_p$ and their incident edges, G' is just a copy of G. Now, pick $f(c) = \tilde{c}$, where $\tilde{c}$ is equivalent to c' restricted to the vertices $V(G) \subset V(G')$. Clearly, $\tilde{c}$ is a coloring of G. We show that it is also a list coloring of G: For a node $v \in V(G)$, assume that $\tilde{c}(v) = i$ and color $i \notin L(v)$. Then c' cannot be a proper coloring of G' since there is an edge $v — v_i$ in G' and $c(v_i) = i$, a contradiction. Since $\tilde{c}$ is a list coloring of G we have $\tilde{c} \in V(K)$.

We show that f is a graph homomorphism as required. Let $\{a, b\} \in E(\mathcal{K}_p(G'))$ and let κ be a Kempe-exchange that is a witness of $a \sim_K b$. There are two cases to consider:

1. The Kempe-exchange κ does not involve any of the nodes $v_1, \dots, v_p$. Then f renames the color classes of the colorings a and b if required and there is a Kempe-exchange corresponding to κ that establishes $f(a) — f(b)$ in K.
2. The Kempe-exchange κ involves two nodes $u, v \in \{v_1, \dots, v_p\}$. We need to consider following two subcases. If $G'(a(u), a(v))$ is connected then $f(a) = f(b)$ and thus, $f(a) — f(b)$ since each node of K has a self-loop. If it is not connected then $f(a)$ and $f(b)$ differ with respect to the color classes $a(u)$ and $a(v)$. We show that $f(a)$ and $f(b)$ are connected by a sequence of Kempe-exchanges that swaps the colors in all except a single connected component of $G'(a(u), a(v))$ and thus $f(a) — f(b)$ by the construction of K above. To obtain $f(b)$, we first apply κ to a on G' and then apply f to the resulting coloring. The Kempe-exchange κ swaps the colors of the connected component of $G'(a(u), a(v))$ containing u and v, and then f swaps the colors of the color classes u and v. As a result, $f(b)$ can be constructed from $f(a)$ by swapping the colors in all connected components of $G'(a(u), a(v))$ except the one containing u and v in the preimage $f^{-1}(V(G(a(u), a(v))))$.

In summary, for all $a, b \in V(\mathcal{K}_p(G')) : a — b$ implies $f(a) — f(b)$. □

The graph homomorphism f defines the following equivalence relation $\sim_f$ on $V(\mathcal{K}_p(G'))$: for $a, b \in V(\mathcal{K}_p(G')) : a \sim_f b$ if $f(a) = f(b)$.

Theorem 2.20 $\mathcal{K}_p(G)[V_\pi]$ *is connected if and only if* $\mathcal{K}_p(G')$ *is connected.*

Proof We noted above that $\mathcal{K}_p(G)[V_\pi]$ is connected if and only if K is connected. Let $f : \mathcal{K}_p(G') \to K$ be the graph homomorphism from Lemma 2.19.

"Only if" part: Let $\mathcal{K}_p(G')$ be connected. Then K is connected since there is a graph homomorphism $\mathcal{K}_p(G') \to K$ and graph homomorphisms preserve connectedness. Thus, $\mathcal{K}_p(G)[V_\pi]$ is connected.
"If" part: Let $\mathcal{K}_p(G)[V_\pi]$ be connected. Then K is connected. Due to the first isomorphism theorem, $K \cong \mathcal{K}_p(G')_{/\sim_f}$ and thus, $\mathcal{K}_p(G')_{/\sim_f}$ is also connected. Any two colorings u, v of G' such that $u \sim_f v$ are connected by Kempe-exchanges since one can be obtained from the other by permuting the colors of the color classes. □

For general graphs, not much is known about the diameter of the corresponding Kempe-k-coloring graphs. Using the graph homomorphism from Lemma 2.19, we show that the reduction from list to graph coloring increases the (possibly unknown) diameter only moderately:

Theorem 2.21 $\operatorname{diam}(\mathcal{K}_p(G)[V_\pi]) \leq \lfloor \frac{|V(G)|-1}{2} \rfloor \cdot \operatorname{diam}(\mathcal{K}_p(G'))$.

Proof For any adjacent nodes u, v in $\mathcal{K}_p(G')$, we count how many Kempe-exchanges are required to get from $f(u)$ to $f(v)$ in $\mathcal{K}_p(G)[V_\pi]$. Let κ be the Kempe-exchange that is a witness of $u — v$, and let i and j be the involved color classes. If $u \sim_f v$ then, in the worst case, all except one connected component of $G(i, j)$ need to be switched to get from u to v for the reasons stated cases 1 and 2 in the proof of Lemma 2.19. There are at most $\lfloor (|V(G)| - 1)/2 \rfloor$ components and one Kempe-exchange is required for each of them. If $u \not\sim_f v$ then there is a single Kempe-exchange on G that establishes $f(u) — f(v)$. Thus, a shortest path of maximum length t in $\mathcal{K}_p(G')$ corresponds to a path of length at most $t \cdot \lfloor (|V(G)| - 1)/2 \rfloor$ in $\mathcal{K}_p(G)[V_\pi]$. □

Given two colorings c, c' of G', we can use the algorithm KEMPERECOLOR (see Algorithm 1) to transform c into c' as long as there is a sufficient number of colors available. However, since G' contains K_p as a subgraph it has degeneracy at least p. According to Theorem 2.10, at least $p + 1$ colors are needed by KEMPERECOLOR and thus, we cannot simply use Theorem 2.10 to prove the connectedness of the search space in the presence of availability requirements. To improve the situation, we assume that the vertices $v_1, \ldots, v_p$ of G' have colors $1, \ldots, p$ for all colorings of interest. Under this assumption, we need to determine a vertex ordering such that KEMPERECOLOR requires fewer colors than the naïve approach. The idea is to use KEMPERECOLOR to generate paths in $\mathcal{K}_p(G')$ such that the algorithm never needs to pick alternative colors for the clique vertices.

In the following, let $C, D \subseteq V(G)$ such that $C \cap D = \emptyset$ and $C \cup D = V(G)$. Further, let $S(G)$ be the set of vertex orderings of G and let $S' \subset S(G)$ be the vertex orderings satisfying

$$\forall u, v \in D, w \in C : u < v \wedge u — v \wedge v — w \Rightarrow w < v \ . \tag{2.3}$$

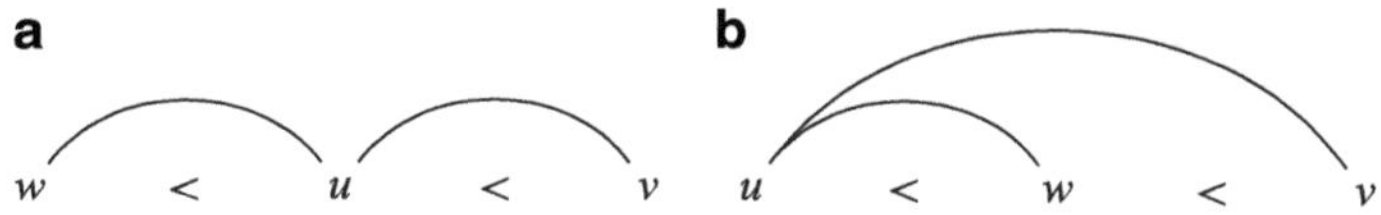

Fig. 2.5 Two vertex orderings of the graph P_3. (**a**) Ordering compatible with Eq. (2.3). (**b**) Ordering incompatible with Eq. (2.3)

That is, if v is a successor of u and they are adjacent, then all neighbors of v in C must precede v. Figure 2.5 shows two examples of vertex orderings of the graph $P_3 = u — v — w$. For $D = \{u, v\}$, $C = \{w\}$, ordering Fig. 2.5a satisfies the condition in Eq. (2.3) and Fig. 2.5b does not.

For $\sigma \in S$, $v \in V(G)$, let $\text{pred}(v, \sigma)$ be the number of predecessors of v adjacent to v. Let the *subdegeneracy* $\lambda(C, G)$ of G relative to C be

$$\lambda(C, G) = \min_{\sigma \in S'} \max_{v \in D} \text{pred}(v, \sigma) \ .$$

Note that $\lambda(C, G) = \deg(G)$ if C is empty. If C is not empty then $\lambda(C, G) \leq \deg(G)$.

Theorem 2.22 *Let c, c' be k-colorings of G that agree on C.* KEMPERECOLOR *transforms c into c' such that all intermediate colorings also agree on C, if the vertices are processed according to $\sigma \in S'$ and $k \geq \lambda(C, G) + 1$.*

Proof We first show that the colors of the vertices C are not changed by KEMPERECOLOR. Assume for a contradiction that in some intermediate coloring a vertex $w \in C$ has a color different from $c(w)$. Then w has been recolored because a neighbor u of w preceding it in σ received color $c(w)$. There are two possible reasons: Either u was recolored to $c(w)$ because $c'(u) = c(w)$, but then $c'(w) \neq c(w)$, a contradiction. If this is not the case, then u was recolored in case 1 or 2 of KEMPERECOLOR, because of a neighbor v preceding it. But this is a contradiction to $\sigma \in S'$.

We now show that $\lambda(C, G) + 1$ colors are sufficient. Since the vertices in C are never recolored, we consider only the vertices D. An unused color may be picked for a vertex $v \in D$ in case 2 of Algorithm 1. There are at most $\lambda(C, G)$ neighbors of v preceding it, and there are at most $\lambda(C, G) - 1$ colors different from the color of v present among these vertices. Thus, there is at least one other color available for v. □

We propose a heuristic approach to finding a witness vertex ordering of $\lambda(C, G)$. Let $\tilde{S} \subset S'$ be the vertex orderings such that the vertices C precede all other vertices. Recall that for any graph G a witness vertex ordering of the degeneracy $\deg(G)$ can be found by repeatedly removing vertices of minimal degree [Mat68, SW68]. In a similar fashion, we can determine an ordering of the vertices D that minimizes $\max_{v \in D} \text{pred}(v, \sigma)$ over all vertex orderings $\sigma \in \tilde{S}$.

Algorithm 3: VERTEXELIMINATION

input : graph G, vertices $C \subseteq V(G)$
output: ordering $v_1, \ldots, v_{|D|}$ of the vertices $D = V(G) \setminus C$

```
G_|D| ⟵ G
for i ⟵ |D| downto 1:
    choose v_i from argmin_{v∈D} {δ(v, G_i)}
    G_{i−1} ⟵ G_i − v_i .
return v_1, ..., v_|D|
```

Theorem 2.23 *The output of* VERTEXELIMINATION *is a vertex ordering which minimizes* $\max_{v \in D} \operatorname{pred}(v, \sigma)$ *over all vertex orderings* $\sigma \in \tilde{S}$.

Proof The proof is based on the remark on the optimality of the vertex elimination algorithm in [Mat68]. Let $\ell = |D|$ and for an ordering $v_1, \ldots, v_\ell$ of D let $G_i = G[C \cup \{v_1, \ldots, v_\ell\}]$. Further, let

$$\hat{\delta} := \max_{G[C] \subseteq H \subseteq G} \; \min_{v \in V(H) \setminus C} \{d(v, H)\} \; .$$

Intuitively, $\hat{\delta}$ is analogous to the degeneracy of G, but the vertices C are irrelevant. If an ordering $\sigma = v_1, \ldots, v_\ell$ of D is an output of VERTEXELIMINATION then

$$\begin{aligned} \max_{1 \le i \le \ell} \operatorname{pred}(v_i, \sigma) &= \max_{1 \le i \le \ell} \{d(v_i, G_i)\} \\ &= \max_{1 \le i \le \ell} \; \min_{v \in V(G_i) \setminus C} \{d(v, G_i)\} \le \hat{\delta} \; . \end{aligned}$$

The graphs G_i coincide with those in Algorithm 3.

Now let H^* be a graph such that $G[C] \subseteq H^* \subseteq G$ and

$$\min_{v \in V(H^*) \setminus C} \{d(v, H)\} = \hat{\delta} \; .$$

Let $v_1, \ldots, v_\ell$ be any ordering of D and let i be the smallest index such that $H^* \subseteq G_i$. Then v_i must be a vertex of H^* and $d(v_i, G_i) \ge \hat{\delta}$. Therefore, for any ordering $v_1, \ldots, v_\ell$ of D, $\max_{1 \le j \le \ell}\{d(v_j, G_j)\} \ge \hat{\delta}$, with equality if the vertex ordering is an output of VERTEXELIMINATION. □

As a potential improvement, the following post-processing step can be performed: Let $v_1, \ldots, v_{|D|}$ be an output of VERTEXELIMINATION and let k be the largest number such that $v_1, \ldots, v_k$ are independent. Then the vertices $v_1, \ldots, v_k$ can be moved before the vertices C in the ordering without violating condition (2.3). The resulting ordering $\sigma' \in S'$ is not in $\tilde{S}$ and can thus not be generated by VERTEXELIMINATION. There is a potential advantage because the construction guarantees that $\max_{v \in D} \operatorname{pred}(v, \sigma') \le \max_{v \in D} \operatorname{pred}(v, \sigma)$.

In summary, the heuristic for computing a vertex ordering σ of G such that $\max_{v \in D} \text{pred}(v, \sigma)$ is close to $\lambda(C, G)$ performs the following two steps:

1. Run VERTEXELIMINATION to generate an ordering $v_1, \ldots, v_{|D|}$ of the vertices D.
2. Let $k \in \mathbb{N}$ be the largest number such that $v_1, \ldots, v_k$ are independent in G. Move the vertices $v_1, \ldots, v_k$ before the vertices C in the ordering.

We apply this heuristic to prove the connectedness of the clash-free timetables that satisfy the availability constraint for a number of benchmark instances. First, we use the reduction from list to graph coloring described above to construct from a conflict graph G the graph G', which contains a clique $v_1, \ldots, v_p$. Then we chose $C \subseteq V(G')$ to include every vertex with $p-1$ neighbors in $\{v_1, \ldots, v_p\}$, that is

$$C = \{v \in V(G') \mid |\Gamma(v) \cap \{v_1, \ldots, v_p\}| = p - 1\} \ .$$

Now we can apply the heuristic to obtain an upper bound $\lambda' \geq \lambda(C, G')$ and check if $p \geq \lambda' + 1$. If this is the case, then Theorem 2.22 implies that the search space is connected.

Results

Table 2.1 shows some results of applying the search space analysis to a number of UCTP benchmarking instances. All instances can be obtained from the SaTT group website, University of Udine [DGS]. The instances `comp01`,...,`comp21` are from the Curriculum-based Course Timetabling (CB-CTT) track of the ITC2007 competition. The instances `ITC2_i01`,...,`ITC2_i24` are from the Post-enrollment Course Timetabling (PE-CTT) track of the same competition. The `erlangen` instances are from the school of engineering at the University of Erlangen-Nürnberg. The `toy` instance is a small example instance from the website [DGS]. The table contains for each instance the number $|P|$ of timeslots, the degeneracy $\deg(G)$ of the conflict graph G, and an upper bound $\lambda'(C, G')$ on $\lambda(C, G')$ produced by the heuristic, where G' is the graph obtained from G by the list coloring reduction and C is set as described above. Entries marked in bold face indicate the connectedness of the corresponding reconfiguration graphs, which is established by Theorems 2.10 (no availability requirements) and 2.22 (with availability requirements).

By Theorem 2.10, reconfiguration graphs of clash-free timetables are connected if $p > \deg(G)$ and by Theorem 2.22, the reconfiguration graphs of clash-free timetables satisfying availability requirements are connected if $p > \lambda(C, G')$ for a suitably chosen $C \subseteq V(G')$. The corresponding values of $\deg(G)$ and $\lambda(C, G')$ in Table 2.1 are marked in bold face in these cases. The data indicates that the clash-free timetables for all CB-CTT and `erlangen` instances are connected, while the requirements of Theorem 2.10 are not satisfied for any of the PE-CTT instances. For eight CB-CTT instances, the upper bound on $\lambda(C, G')$ is sufficient to show that the reconfiguration graphs are connected in the presence of availability

Table 2.1 The number of timeslots, deg(G) and upper bounds $\lambda'(C, G') \geq \lambda(C, G')$ produced by the heuristic are given for each instance

Instance	$\vert P\vert$	deg(G)	$\lambda'(C, G')$	Instance	$P\vert$	deg(G)	$\lambda'(C, G')$
`comp01`	30	**23**	**24**	`ITC2_i01`	45	91	109
`comp02`	25	**23**	30	`ITC2_i02`	45	99	119
`comp03`	25	**22**	27	`ITC2_i03`	45	73	92
`comp04`	25	**17**	25	`ITC2_i04`	45	78	100
`comp05`	36	**26**	43	`ITC2_i05`	45	81	99
`comp06`	25	**17**	28	`ITC2_i06`	45	80	100
`comp07`	25	**20**	**24**	`ITC2_i07`	45	80	106
`comp08`	25	**20**	**24**	`ITC2_i08`	45	69	97
`comp09`	25	**22**	25	`ITC2_i09`	45	89	108
`comp10`	25	**18**	27	`ITC2_i10`	45	97	116
`comp11`	45	**27**	**27**	`ITC2_i11`	45	75	93
`comp12`	36	**22**	40	`ITC2_i12`	45	91	109
`comp13`	25	**17**	**22**	`ITC2_i13`	45	87	106
`comp14`	25	**17**	**23**	`ITC2_i14`	45	87	107
`comp15`	25	**22**	27	`ITC2_i15`	45	79	106
`comp16`	25	**18**	25	`ITC2_i16`	45	55	83
`comp17`	25	**17**	25	`ITC2_i17`	45	50	71
`comp18`	36	**14**	**32**	`ITC2_i18`	45	91	112
`comp19`	25	**23**	27	`ITC2_i19`	45	101	120
`comp20`	25	**19**	**23**	`ITC2_i20`	45	73	92
`comp21`	25	**23**	28	`ITC2_i21`	45	72	90
`erl.2011-2`	30	**22**	32	`ITC2_i22`	45	98	118
`erl.2012-1`	30	**14**	31	`ITC2_i23`	45	117	128
`erl.2012-2`	30	**20**	32	`ITC2_i24`	45	77	97
`erl.2013-1`	30	**16**	30	`toy`	20	**10**	**11**

For each instance, G' is obtained from the conflict graph G by the reduction from graph to list coloring. The instances are the CB-CTT, PE-CTT and Erlangen benchmarking instances from [DGS]. Entries marked in bold face indicate the connectedness of the corresponding reconfiguration graphs

constraints. The situation is quite different for the PE-CTT instances, since neither deg(G) nor $\lambda'(C, G')$ is sufficient to show the connectedness of the reconfiguration graphs. Getting better bounds on $\lambda(C, G')$ is of no use since $\lambda(C, G') \geq \deg(G)$. Thus, a new approach for proving the connectedness or disconnectedness of the reconfiguration graphs is likely to be required for these instances. In Appendix A.1, the degeneracy values of the conflict graphs are documented for other instance sets without timeslot availability requirements.

We will show that for the instance `toy`, the proposed heuristic yields a vertex ordering that is a witness for the value $\lambda(C, G')$. Let G be the conflict graph for this instance and let G' be the graph that results from the reduction from list to graph coloring. In the CB-CTT formulation the events are grouped into courses and

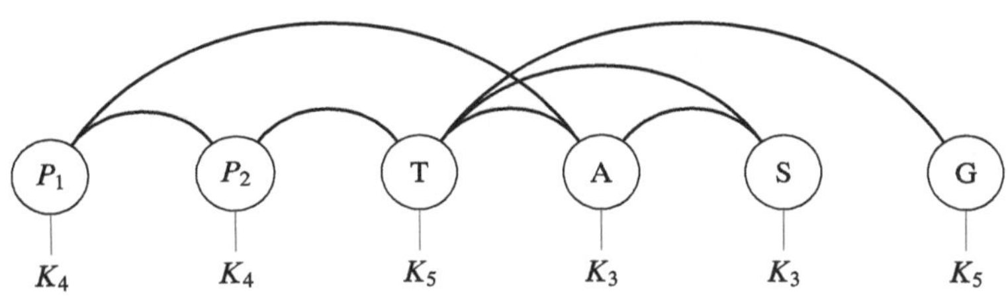

Fig. 2.6 Succinct representation of an optimal vertex ordering of the graph G' obtained from the conflict graph of the instance `toy` by the reduction to graph coloring. All nodes represent cliques as denoted below the nodes

for each course, events of the course are a clique in G and G'. Similarly, if two courses are in conflict, then the events of both courses are a clique in G and G'. If certain timeslots are unavailable for a course, then the events of the course and the timeslots form a clique in G'. In the `toy` instance, there are four courses which consist of 16 events in total. Figure 2.6 shows a succinct representation of an optimal vertex ordering of the graph G'. Each node of the shown graph is a clique, as noted below the nodes, and the cliques are ordered from left to right. Two nodes of the shown graph are connected if all nodes of the corresponding cliques are connected. The nodes T, A, S and G correspond to the courses labeled SceCosC, ArcTec, TecCos and Geotec, respectively. The node P_1 represents to the timeslots marked unavailable for course ArcTec and the node P_2 represents the timeslots unavailable for SceCosC. Any two conflicting courses are connected. Let $C = V(P_1) \cup V(P_2)$.

Let σ be a vertex ordering of G' such that the cliques are arranged in the order P_1, P_2, T, A, S, G with some arbitrary choice of the relative ordering of the vertices within each clique. This ordering is a possible output of the algorithm VERTEX-ELIMINATION. From

$$\max_{v \in V(G') \setminus C} \operatorname{pred}(v, \sigma) = 11 \ ,$$

we can conclude that $\lambda(C, G') \leq 11$.

Proposition 2.24 *For the instance* `toy`, $\lambda(C, G') = 11$.

Proof Let σ be an ordering of $V(G')$ and $V' \subseteq V(G') \setminus C$. The maximum number of predecessors adjacent to any vertex of V' in G' is denoted by

$$p(V', \sigma) = \max_{v \in V'} \operatorname{pred}(v, \sigma) \ .$$

Note that for a clique $K \in \{T, A, S, G\}$, the value $p(K, \sigma)$ is determined by the last vertex of K in σ. Thus, the value of $p(K, \sigma)$ depends only on the relative order of the last vertices of the cliques $\{T, A, S, G\}$ in σ. Let $\tilde{S}$ be the vertex orderings of G' such the vertices C precede all other vertices of G' and let $\hat{S}$ be the total orderings of $\{T, A, S, G\}$. For each ordering $\sigma' \in \hat{S}$ we can pick an ordering $\ell(\sigma')$ of G' that

Table 2.2 Enumeration of the maximum number $p(V, \sigma)$ of adjacent predecessors of the vertices V with respect to the ordering σ for all relevant cases in the proof of Proposition 2.24

Clique ordering $\sigma' \in \hat{S}$	$p(T, \ell(\sigma'))$	$p(A, \ell(\sigma'))$	$p(S, \ell(\sigma'))$	$p(G, \ell(\sigma'))$
T, A, S, G	8	11	10	9
T, S, A, G	8	14	7	9
A, T, S, G	11	6	10	9
S, T, A, G	11	11	2	9
A, S, T, G	14	6	5	9
S, A, T, G	14	9	2	9

is compatible with σ' in the sense that the relative ordering of the last vertices of the cliques is in accordance with σ'. We have,

$$\lambda(C, G') = \min_{\sigma \in \tilde{S}} \max_{K \in \{T,A,S,G\}} p(K, \sigma) = \min_{\sigma' \in \hat{S}} \max_{K \in \{T,A,S,G\}} p(K, \ell(\sigma')) \ .$$

For any ordering $\sigma' \in \hat{S}$ such that $G < T$, we have $p(T, \ell(\sigma')) \geq 13$, because the last vertex of T has at least 13 adjacent predecessors in G'. Thus, we only need to consider orderings such that $G > T$. Furthermore, since no vertex of G is adjacent to any vertex of A or S, changing the relative order of A and G or S and G does not change the number of adjacent predecessors. Hence, we can assume G is a maximum in any ordering of interest and therefore only the six cases remain, which we deal with by enumeration. In Table 2.2 we give the values of $p(K, \ell(\sigma'))$ all for $K \in \{T, A, S, G\}$ and all permutations of $\{T, A, S\}$. From these values we get

$$\lambda(C, G') = \min_{\sigma' \in \hat{S}} \max_{K \in \{T,A,S,G\}} p(K, \ell(\sigma')) = 11 \ ,$$

which concludes the proof. □

From Proposition 2.24 we can conclude that the proposed heuristic produces a witness of $\lambda(C, G') = 11$ on the instance `toy`.

2.4 Solution Approaches

In this section we will classify and review algorithms for solving the UCTP search problem. A solution to the search problem is a by-product of solving the optimization problem, so this is in itself not a restriction. However, we are particularly interested in cases in which the search problem itself is challenging and thus receives special attention. After reviewing different approaches from the literature, we will discuss the Kempe Insertion Heuristic (KIH) proposed in [MW10b*]. Then, we will give a SAT formulation of the UCTP, which we will use to provide a point of reference in our experimental evaluation of the KIH in addition to the results generated by the heuristic state-of-the-art approaches.

Many of the heuristic algorithms for the UCTP take at some point a constructive approach, that is, the events are inserted into the timetable one-by-one such that clash and overlap freeness as well as other requirements are maintained. Thus, at some point during the construction, a subset $E' \subseteq E$ has been assigned to resources while the remaining events are yet unassigned. In these cases, the timetable $\tau : E \to P \times R$ is a partial function and to make clear the distinction we call it a *partial timetable*. We refer to the number of unscheduled events in a partial timetable as the *distance to feasibility*. The distance to feasibility of a feasible timetable is zero.

2.4.1 Existing Approaches

In this section, we provide an overview of the approaches in the literature targeted at solving the search version of the UCTP. Many techniques such as the sequential heuristics have been studied extensively in the context of examination timetabling [CLL96,QBM+09]. Due to the similar structure of the UCTP and examination timetabling problems, the techniques developed for examination timetabling problems generally carry over to the UCTP and vice versa. Lewis and Paechter argued that finding feasible timetables is too easy on existing benchmark sets, because they were intended not as search problems but as optimization problems with a focus on minimizing soft constraint violations [LP07]. Thus, they proposed a set of 60 artificially generated instances which are challenging with respect to finding feasible timetables. We will restrict our literature review to four categories of approaches that are relevant for this instance set, namely sequential heuristics, Local Search (LS), population-based heuristics, and hyper-heuristics. For a complete overview of the vast body of literature on algorithms for the UCTP, please refer to the very good surveys [CL96, Sch99, BdWK03, QBM+09].

Sequential Heuristics

Sequential heuristics assign events to resources one by one and are in essence adaptations of the greedy coloring algorithm to the UCTP. The algorithmic scheme works as follows: The input is a UCTP instance, and there are two policies: policy P_1 that picks the next event to be scheduled, and a policy P_2 that selects a resource for a particular event. In each iteration, the sequential heuristic picks an event e according to P_1 and assigns an available resource to e according to P_2. If P_1 depends on the events that have already been scheduled, the sequential heuristic is called *dynamic*. Otherwise, the sequential heuristic is called *static*, and the order in which events are scheduled is determined a priori. The policy P_2 typically chooses the first suitable resource from a sequence of candidate resources. For instance, in [TBM07, MW10b*], the resources are arranged in random order and the first suitable resource is chosen. We will refer to this resource selection policy as Random Ordering (RO) policy. Additional constraints may be considered by P_2 when choosing one of several available resources [CLL96]. If at some point there are no available resources for the current event, it is left unassigned and the resulting timetable will not be feasible. Choices of P_1 from the literature [CL96, BdWK03] include:

- *Largest Degree (LD) (static)*: The events are arranged in non-increasing order according to their degrees in the conflict graph, see [WP67]. The next event is selected according to this ordering.
- *Largest Weighted Degree (LWD) (static)*: For each event, the total number of students involved in its conflicts is determined and the events are arranged in non-increasing order according to these values. The next event is selected according to this ordering.
- *Random (static)*: The next event is picked at random from the remaining events.
- *Least Saturation Degree (LSD) (dynamic)*: An event with the least number of available resources is selected as the next event to be scheduled.
- *Largest Color Degree (LCD) (dynamic)*: An event with the largest number of conflicts in the current timetable is selected as the next event to be scheduled.

Since sequential heuristics are computationally cheap, they can provide a starting point for computationally more involved heuristics [MW10b*, TBM07].

Local Search-Based Heuristics

Local Search (LS) is a class of heuristics for solving combinatorial search and optimization problems. Popular classes of LS-based heuristics are

- Simulated Annealing (SA) [KGV83],
- Tabu Search (TS) [GL97],
- Hill Climbing (HC) [BB12], and
- Great Deluge (GD) [Due93].

The general scheme fits quite well in the formal framework of reconfiguration problems from Sect. 2.3.1. Given some adjacency relation on the set of solutions and an initial solution, the search space is traversed by iteratively moving to neighboring solutions. LS-based heuristics differ with respect to the adjacency relation and the policy that determines under which conditions the current solution is replaced by a neighbor, and which of the neighbors is chosen to replace the current solution. Several adjacency relations may be used individually or in combination by a LS-heuristic [DGS06].

Virtually all neighborhood structures that appear in LS-based algorithms for the UCTP and related problems are based on the Kempe-exchange, which is a popular operation used in various solvers for timetabling problems over the last two decades [TD98, MBHS02, LH10, TBM07]. We will present a unified view on the neighborhood structures in terms of which subsets of Kempe-exchanges are considered. In addition to choosing which Kempe-exchanges may be performed to generate a neighborhood, there are a variety of options how to deal with availability requirements and the room assignment when moving events between timeslots.

- Move Event Neighborhood (ME): For a given event e the neighborhood contains all timetables that result from moving e to a different unoccupied resource. Moving an event e in this is equivalent to a Kempe-exchange that operates on the Kempe-component $(\{e\}, \emptyset)$.
- Swap Event Neighborhood (SE): Contains all timetables that result from swapping the timeslots of exactly two events. A swap of two events e_1 and e_2 is equivalent to a single Kempe-exchange that operates on the Kempe-component $(\{e_1, e_2\}, \{e_1 — e_2\})$ if the two events are conflicting, or two consecutive steps in the ME neighborhood otherwise.
- Kempe-exchange Neighborhood (KX): For a given event e, the neighborhood contains all timetables that result from applying a Kempe-exchange such that e is contained in the Kempe-component. This neighborhood has been used in the context of examination timetabling [MBHS02, BEM+10].

Typically, only such neighbors are considered that respect overlap-freeness and availability requirements [TBM07, CDGS12, MBHS02, BEM+10]. Generating neighbors in the KX neighborhood while respecting room availability requirements is computationally expensive in comparison to the simpler ME or SE neighborhoods: the former requires solving two room assignment subproblems, i. e. two maximum cardinality matchings, to determine a feasible room assignment for all events.

We will discuss two local search approaches that are considered state-of-the art in more detail. The general idea that is common to both is that the number of events that cannot be scheduled in the timetable without violating overlap freeness and availability requirements is minimized. In both approaches, there seemed to be some gain in minimizing the number of students attending the events that were not properly scheduled instead of just counting the number of such events.

Hybrid Simulated Annealing [TBM07] The hybrid simulated annealing algorithm uses three neighborhood structures, ME, SE, and a neighborhood based on a special variant of the Kempe-exchange, called the Pair-wise Kempe Chain (PKC) neighborhood. A move to a neighbor solution in the SE neighborhood exchanges the resource assignments of two events. The pair-wise Kempe-exchange works as follows: Let G be the conflict graph of a UCTP instance, τ be a partial timetable and p_1 and p_2 be two distinct timeslots. Add an edge between two nodes of G whenever the corresponding events are assigned to the same room according to τ. The pairwise Kempe-exchange exchanges the resources assigned to any two events in a Kempe-component of $G(p_1, p_2)$ that are scheduled in the same room. Any pairwise Kempe-exchange operates on a union of ordinary Kempe-components and is therefore equivalent to performing one or more Kempe-exchanges. However, the benefit of the pair-wise Kempe-exchange is that assigning rooms for all events in a

Kempe-component is straight-forward due to pairing events in the same room in the two timeslots.

In a preprocessing step, the algorithm uses a variant of the sequential heuristics: In each step, the heuristic chooses the next event to be scheduled according to the least saturation degree and ties are broken according to the largest degree. Each event is assigned to a random suitable resource. Events that have no suitable resources left are accommodated in artificial timeslots, such that conflict and overlap requirements are satisfied. The whole preprocessing step consists of 500 trials of this sequential heuristic. After the preprocessing, the number of artificial timeslots is minimized using an SA-based approach. For each temperature level, the algorithm first searches the neighborhood ME followed by the neighborhood SE, accepting neighbors according to the usual probabilistic SA acceptance criterion, see [KGV83]. If no improvement has been made for a fixed number of trials in these neighborhoods, the algorithm switches to the PKC neighborhood. Again, if no improvement has been made within a fixed number of iterations, the algorithm switches back to the other two neighborhoods after adjusting the temperature. Several steps in the PKC neighborhood may be performed before the resulting timetable is evaluated and compared to the prior timetable. The cooling schedule decreases the temperature by the factor $T/(1 + T\beta)$ when the temperature is adjusted, where β is an algorithm parameter that is picked from the interval $[0.0005, 0.001]$.

Tuned Simulated Annealing [CDGS12] This approach includes the ME and SE neighborhoods in a fairly standard SA setting that is tuned to the UCTP search problem (and several related optimization problems) using statistical methods. Room and timeslot availability requirements as well as overlap-freeness requirements are satisfied by design. In contrast to the hybrid simulated annealing proposed in [TBM07], conflicts are tolerated and minimizing the number of event conflicts is part of the objective. The room assignment possibilities in the SE neighborhood are more relaxed in comparison to the SE neighborhood in [TBM07]: whenever two events in different timeslots are swapped, any available, unoccupied in the respective target timeslot can be assigned. Similar to [TBM07], there is an artificial timeslot that accommodates the events not yet scheduled. There is an additional restricted neighborhood, ME^{-}, that prohibits moving events to the artificial timeslot.

Some constraint propagation is performed in a preprocessing step. For instance if two events can be scheduled only is the same single room they cannot be scheduled in the same timeslot and are thus marked as conflicting. After this step, one of two possible construction procedures is performed to insert events into the timetable. The first procedure, referred to as I_0 picks for each event e a random timeslot and schedules the event in this timeslot if an unoccupied room is available. If none is available, the event is scheduled in the artificial timeslot. The second procedure, I_1 tries for each event to find an unoccupied room in a number of available timeslots. After the constructive procedure, the simulated annealing is performed. The cooling scheduled is geometric, i. e., in each temperature update, the temperature is multiplied by a decay factor $\beta \in (0, 1)$. The objective is to minimize the sum of the event conflicts and the number of students in the events scheduled in the artificial timeslot.

In order to tune the general algorithmic scheme to the instances at hand, three different combinations of neighborhoods and constructive procedures were evaluated and for these three, the parameter space was sampled using Nearly Orthogonal Latin Hypercubes (NOLH) [CL07]. The best configurations were then determined by the F-race algorithm [BYBS10, Bir], see also section "Preliminaries: Algorithm Configuration". The tuning process was performed separately for the various sets of instances, since the problem formulations differ between different instance sets. For a comparison with the KIH from Sect. 2.4.2, the most relevant instances are those generated by Lewis and Paechter [LP07, LPb]. The best configuration on these instances was found to a combination of initialization I_0 followed by the SA procedure using the union of the neighborhoods ME and SE. The initial temperature was set to 31.62 on the medium instances and set slightly higher, to 36.30, on the big instances. The minimum temperature was determined to be approximately 0.12 for both instance sets. The number of neighborhood searches per iteration was configures such that the minimum temperature was reached after a given number of iterations with $\beta = 0.9999$.

Population-Based Heuristics

Population-based heuristics include Genetic Algorithms (GA), Evolutionary Algorithm (EA), Particle Swarm Optimization (PSO), and other nature-inspired optimization techniques [BFM97, Ken10, DS04]. In contrast to LS-based methods, a population of candidate solutions is maintained, which are modified iteratively to identify and explore promising regions of the search space. A key difference to other approaches is that there is some mechanism for exchanging information between different candidate solutions in the population. A brief overview over the EA is given in Sect. A.2.

Grouping Genetic Algorithm (GGA) [LP07] Lewis and Paechter apply a GGA to the UCTP. Following the work of Falkenauer [Fal94], their motivation is that traditional solution representations in genetic algorithms include too much redundancy and thus the traditional mutation and crossover operators do not perform well. A grouping problem captures the task of gathering groups of objects such that a certain objective function is minimized. Among other combinatorial problems, the UCTP can be described as a grouping problem: events need to be grouped in timeslots such that clash and overlap freeness requirements are satisfied. Similar to the local search approach in [TBM07], artificial timeslot are introduced to accommodate events that cannot be scheduled in any of the regular timeslots without violating conflict or overlap freeness requirements.

The initial population is generated using a constructive mechanism that combines the saturation degree, largest degree and random sequential heuristics with three different policies for choosing rooms. Artificial timeslots are added to a partial solution as needed. The mutation operator deletes groups (timeslots) from a candidate timetable and inserts all events that were assigned to these timeslots using the aforementioned constructive mechanism. The crossover operator follows the approach for grouping problems proposed by Falkenauer [Fal94, Fal98]. Given two parents s_1 and s_2 an offspring is generated by the crossover operator in the following way: A set of consecutive group is chosen from s_1 and is injected after a randomly chosen groups in s_2. The standard approach [Fal94] suggests to remove events that occur twice from the groups they were part of originally in s_2. Lewis and Paechter argue that this does not work well for the UCTP since the number of timeslots is likely to be increased by the crossover operation in comparison to the parents and suggest to remove all groups of s_2 that contain events injected by the crossover. The events that were removed from the timetable altogether are then inserted using the constructive mechanism in the same way as the mutation operator. In addition, a local search procedure is added in order to improve the overall performance of the GGA. During the local search, two tasks are performed: First, for each of the unoccupied resources, the algorithm checks if any of the unplaced events can be assigned to them. If this is the case, the event is assigned to the resource. In the second step, a number of moves are performed in the ME neighborhood.

The impact of several objective functions on the performance of the GGA (with and without local search) has been investigated in [LP07]. The overall results shown on the website [LPa] were generated using the objective function

$$f(s) = \frac{\sum_{i=1}^{g} C_i^2}{g} ,$$

proposed by Erben in [Erb01], where s is a timetable in the GGA representation containing groups $1, \ldots, g$ of events and C_i is the number of conflicts of the events in group i with other events in the timetable. This objective function is to be maximized and rewards grouping events of high degree.

Hyper Heuristics

A hyper heuristic solves search and optimization problems by selecting heuristics from a pool of *low-level heuristics* applicable to a concrete problem domain. The aim of a hyper heuristic is maximal generality in the sense that it can work on any problem domain as long as it is provided with a set of suitable low-level heuristics, which encapsulate all problem-dependent knowledge. A crucial component of a hyper heuristic approach is the *high-level heuristic* which identifies the most successful low-level heuristics for a given problem either during the optimization process or offline [BLSS05].

Hybrid Graph-Based Hyper Heuristic (HGHH) [BMM+07] The HGHH uses TS to find a schedule of the sequential heuristics discussed above. The schedule is a list of sequential heuristics that determines which heuristic is applied to insert an event in the timetable. The neighborhood consists of all heuristic schedules that can be generated from the current schedule by exchanging two heuristics. In addition to the tabu list the algorithm keeps track of prefixes of schedules that produce infeasible timetables. If a new schedule matches any of the prefixes it is discarded immediately. Furthermore, each entry in the heuristic schedule is responsible for inserting two events in the timetable, which considerably reduces the size of the search space. For various Examination Timetabling (ETT) instances, Burke et al. report an improvement over using the sequential heuristics individually, even if additional backtracking mechanisms are employed, as in [CL96]. Hybridisation schemes that combine various alternative local search variants as high-level heuristics with the sequential low-level heuristics have been investigated in [QB08]. The results indicate that Iterated Local Search (ILS) and Variable Neighborhood Search (VNS) are to be preferred over TS.

Other Approaches

There are certainly approaches that do not clearly follow any of the schemes discussed above, for example Ant Colony Optimization (ACO) [NMCR12]. The following rather recent approach is roughly based on the construction versus perturbation dichotomy, similar to the Kempe Insertion Heuristic discussed in Sect. 2.4.2:

Clique-Based Construction [LZC11] At the heart of the clique-based, or rather independent set-based, algorithm lies the task of finding large independent sets in the conflict graph G of a UCTP instance. Liu, Zhang, and Chin rely on a heuristic for maximum clique problem proposed in [Öst02], and apply the algorithm on the complement $\overline{G}$ of the conflict graph. For performance reasons, the backtracking part of the heuristic is left out and what remains is the following procedure: Let $c \subseteq V(\overline{G})$ be a clique, possibly $c = \emptyset$. Let $U = V(G)$ if c is empty and $U = \mathrm{N}_{\overline{G}}(c)$ otherwise. While U is not empty, add a vertex v of maximum degree in $\overline{G}$ to c and replace U by $U \cap \mathrm{N}(v)$. The algorithm terminates when U is empty and at this point c contains a clique that is potentially larger than the original clique. Dually, this algorithm extends a given independent set c in G to a maximal independent set.

The algorithm proceeds in three steps, the initialization phase, the recombination phase and the perturbation phase. After the initialization, the algorithm alternates between recombination and perturbation until a feasible timetable has been found or the timeout has been hit. During the initialization, for each timeslot a maximal independent set is generated, then the room assignment subproblem is solved, see Sect. 2.2. In each iteration of the recombination phase, each event in the timetable is removed with probability ρ, a parameter which deteriorates exponentially over time. Then, for each timeslot p, the maximum independent set heuristic above is applied to extend the current independent events in p to a maximal independent set of events. For a timeslot p, the new independent set is accepted only if it is an improvement in terms of the (i) number of events scheduled in p (ii) number of students in the events scheduled in p (iii) sum of the degrees of the events in p. The rationale behind (iii) is that if the sum of the degrees of the events increases, then more of the "difficult" events could be placed in p, which is desirable because no effort has to be spent on placing these events in other timeslots. The recombination phase terminates whenever ρ is below a given minimum probability or a feasible timetable has been found. The perturbation phase exchanges events between different timeslots. The exchange of events is similar to the Kempe-exchange. The main difference is that only a subset of a Kempe-component is processed. Let e be an event that should be moved from timeslot p_1 to p_2. When e is moved to p_2 all events in p_2 conflicting with e are simultaneously moved to p_1. If any of the latter events are in conflict with any event in p_1 then the move is not performed. Similarly, if there is no feasible room assignment after swapping the events, the move is also not performed. The perturbation phase is finished after a fixed number of iterations.

2.4.2 The Kempe Insertion Heuristic

It can be readily observed that the purely constructive approaches, which insert events in the timetable one-by-one while respecting clash and overlap freeness, are rather unsuccessful on many of the UCTP and ETT standard instances. Typically, at some point during the construction of the timetable there are no suitable resources left for a set of events, which are therefore left "unscheduled". In order to avoid such a situation, backtracking [CLL96] or look-ahead [BN99, BMM+07] strategies have been applied. More recent state-of-the-art approaches relax the requirements in some way, for example, by allowing unscheduled events [Kos05, LZC11, MW10b*], adding extra timeslots [TBM07, CDGS12], or tolerating event conflicts [CDGS12]. The goal is then to minimize the violations of the original requirements. The algorithm KEMPEINSERTION proposed in [MW10b*] relaxes the requirement that all events need to be scheduled, similar to the approaches [Kos05, LZC11]. In order to create opportunity for inserting additional events and ultimately find a feasible timetable, KEMPEINSERTION iteratively modifies partial timetables using Kempe-exchanges. In comparison to the other approaches, there is a significant difference with respect to how opportunity to insert events is generated.

The algorithm KEMPEINSERTION is composed of a preprocessing phase followed by a construction and a perturbation. The latter two are performed alternatingly until a feasible timetable has been found or the time limit has been reached. The purpose of the preprocessing is to find a good starting point for the rest of the KIH with a reasonable computational effort. This is accomplished by a variant of the sequential heuristics discussed in Sect. 2.4.1. During the construction phase, additional events are inserted in the timetable by clearing timeslots from events conflicting with the yet unscheduled events. In contrast to other approaches, we use Kempe-exchanges to deliberately move these conflicting events to other timeslots. Similar to the sequential heuristics, the constructive part of the KIH may get stuck and thus, a perturbation mechanism is needed in order to generate feasible timetables on many benchmark instances. The perturbation inserts yet unscheduled events in the timetable and removes other events from the timetable to ensure that conflict and overlap requirements are respected. A high level view of KEMPEINSERTION is shown in Algorithm 5. We will first discuss the construction and the perturbation phase individually and then show how construction and perturbation were balanced for good performance on the benchmark instances.

Preprocessing

The preprocessing runs a sequential heuristic for a fixed number of iterations and keeps a timetable with the lowest distance to feasibility. Multiple runs of the sequential heuristic are performed in order to compensate for the variance in the distance to feasibility of the generated timetables. The behavior of a sequential heuristic is determined by the event selection policy and the resource selection policy, see Sect. 2.4.1. In [MW10b*], we used as sequential heuristic the one that was advocated in [TBM07] as preprocessing for the Hybrid Simulated Annealing algorithm: The next event is selected according to the LSD policy and ties are broken by the LD policy (see p. 40). We will refer to this event selection policy as LSDLD. Suitable resources are selected at random (policy RO). In our experimental evaluation presented in Sect. 2.5, we will investigate the distance to feasibility attained by different sequential heuristics including the LSDLD/RO heuristic on a number of benchmark instances. As a possible improvement to the RO policy we propose the following resource ordering policy, which we refer to as Saturation Ordering (SO). The SO selection policy arranges the timeslots that do not contain events conflicting with e in non-decreasing order according to how many of the remaining events can be scheduled in the timeslot without violating clash or overlap freeness requirements. Then, for each timeslot p in the given order, the room assignment is performed for all events already in p and e. If there is a room assignment for these events, then e can be inserted in the timetable. Please note that rooms may be re-assigned for the other events in timeslot p. If e cannot be inserted in any of the suitable timeslots in this fashion it is left unassigned.

The Construction Phase

Consider inserting events into a partial timetable using a sequential heuristic. At some point the sequential heuristic may not be able to insert the next event e into the timetable, because for each timeslot p suitable for e, at least one of the following obstructions is present:

1. a conflicting event e' of e is in timeslot p.
2. there is no feasible room assignment for the events in timeslot p when e is scheduled in p.

Instead of using backtracking or lookahead techniques as a remedy, our key observation is that the event e can still be inserted in the timetable, if we manage to clear the obstructions for any of the suitable timeslots by using Kempe-exchanges. For this purpose we consider the following problem:

Definition 2.25 (Event Insertion Problem (EIP))
INSTANCE: A UCTP instance $\mathcal{I}$, a partial timetable τ , an event e and a timeslot p.
TASK: Find a sequence of Kempe-exchanges that clears obstructions 1 and 2 for e and timeslot p.

Clearly, if we manage to solve a suitable EIP instance, we can insert an additional event into the timetable and are one step closer to a feasible solution.

We will first have a look at this problem from the reconfiguration problem perspective. Let $G[\tau]$ denote the subgraph of the conflict graph induced by the events in the partial timetable τ. We are looking for a path in $\mathcal{K}_p(G[\tau])$ from the coloring given by τ to a coloring that does not exhibit obstructions 1 and 2. We cannot assume in general that such a path exists: Let $\ell \geq 3$, $k = \ell + 1$ and let $G \times H$ denote the categorial product of two graphs G and H. We can pick a UCTP instance $\mathcal{I}$ and a partial timetable τ for $\mathcal{I}$ such that $G[\tau] = K_\ell \times K_k$. Assume that τ induces a k-coloring of $G[\tau]$. By [Moh07, Proposition 2.1], $G[\tau]$ is ℓ-colorable and the unique ℓ-coloring is not connected to any k-coloring. Thus, it is impossible to clear obstruction 1 for an event that is to be scheduled in timeslot k. To the best of our knowledge, so far no hardness results are known for the problem of deciding the connectedness of Kempe-k-coloring graphs. Due to these considerations, it is reasonable to make only a limited effort to check if the obstructions can be cleared easily.

The procedure EVENTINSERTION is shown in Algorithm 4. It is used repeatedly in the construction phase of KEMPEINSERTION. The loop in line 1 deals with obstruction 1, while loop 2 deals with obstruction 2. If EVENTINSERTION indicates success, then after performing the Kempe-exchanges returned by the algorithm timeslot p does not contain any events conflicting with e and there is at least one unoccupied room in p. The algorithm attempts to clear events conflicting with e by moving all such events to other timeslots using Kempe-exchanges. The alternative timeslots for the conflicting events are tried in random order. However, such Kempe-exchanges can introduce additional conflicting events to timeslot p. To keep things simple and computationally cheap, only such Kempe-exchanges are considered admissible that do not introduce any events conflicting with e. In order to clear obstruction 2, EVENTINSERTION searches for a Kempe-exchange that does not introduce conflicts of e to p and frees up at least one room. In the presence of room availability requirements, EVENTINSERTION can be modified in a straightforward manner such that a room suitable for e is freed up. In total, EVENTINSERTION performs $O(|R| \cdot |P|)$ Kempe-exchanges in order to clear obstructions 1 and 2.

The loop in line 1 of Algorithm 5 contains the constructive part of the KEMPEINSERTION, which essentially uses EVENTINSERTION to solve the EIP repeatedly for events that are yet to be scheduled. In each iteration, an event e from the list of unscheduled events is picked uniformly at random (uar). Then, the

Algorithm 4: EVENTINSERTION

input : UCTP instance $\mathcal{I}$, partial timetable τ for $\mathcal{I}$, event e, timeslot p
output: a sequence of Kempe-exchanges, indication of success

$K \longleftarrow$ EmptyList
1 **foreach** *conflict e' of e in p*:
    find timeslot $p' \neq p$ such that Kempe-exchange (e', p, p') does not introduce events conflicting with e in p
    if *suitable p' was found*:
        append (e', p, p') to K
    else:
        return K, no success
if *there is a free room for e in p*:
    return K, success
2 **foreach** *event e' in timeslot p*:
    find timeslot $p' \neq p$ such that Kempe-exchange (e', p, p') decreases number of events in p
    if *suitable p' was found*:
        append (e', p, p') to K
        return K, success
return K, no success

timeslots available for e are processed in random order, and for each timeslot p, EVENTINSERTION is employed to solve the EIP instance $(\mathcal{I}, \tau, e, p)$. The Kempe-exchanges returned by EVENTINSERTION are applied to the timetable τ and the room assignment subproblem is solved for the events in timeslot p and e. If room availability requirements are present, it is possible at this point that there is no feasible room assignment, in which case e is not scheduled and we continue with the next timeslot. If the attempt to scheduled e in p was successful and the resulting partial timetable τ has fewer unscheduled events than the current best timetable τ_{best}, the latter is replaced by τ. As shown in line 2 of Algorithm 5, if EVENTINSERTION is unsuccessful for a particular timeslot p, the events conflicting with the event e are collected in p to increase the chances of successfully solving the EIP for other timeslots. This operation basically does the opposite of what EVENTINSERTION does: We try to increase the number of events conflicting with e by performing Kempe-exchanges between p and all other timeslots that are yet to be processed in the current iteration.

The constructive part has been described in [MW10b*] in terms of a neighborhood search that repeatedly samples timetables from a set $\mathcal{N}_{e,p}(\tau)$ of *neighbors*, for a given event e and a timeslot p:

$$\mathcal{N}_{e,p}(\tau) = \{\tau' \mid \tau' \text{ is obtained from } \tau \text{ by solving the EIP } (\mathcal{I}, \tau, e, p)\} .$$

In contrast to other local search approaches to the UCTP and related problems that sample a number of solutions from a neighborhood [DGS01, TBM07], the difficulty here is to generate a single neighbor. Also, there is no immediate use in sampling multiple neighbors and taking the best one, as it is common, for example, in the tabu search metaheuristic [Glo89], since all of them have the same distance to feasibility.

The Perturbation Phase

The purpose of the perturbation phase is to introduce some changes to the list of unscheduled events whenever the constructive phase gets stuck. Thus, there will be different unscheduled events in the next construction phase which can potentially be inserted into the timetable. In essence, the perturbation can undo decisions that were taken earlier in the search process by removing events from the timetable. A perturbation operation consists of picking an unscheduled event e uar. Then the suitable timeslots for e are arranged in non-decreasing order by the number of events they contain that are in conflict with e. A timeslot is then picked from this sequence according to random variable that is exponentially distributed with parameter λ. Due to the characteristics of the distribution, it is more likely to pick a timeslot with few events conflicting with e. Once a timeslot has been picked, all events conflicting with e are removed from p and e is scheduled in the timeslot. In [MW10b*], the perturbation mechanism was termed *forced event insertion*. Typically, when KIH has made some progress, several conflicting events are removed from the timetable during the perturbation, so the distance to feasibility is increased. The aim is to use the perturbation just often enough so that the construction phase can insert additional events. In the original work [MW10b*], we used a scheme based on counting the number of successful attempts to insert events into the timetable to balance the usage of the constructive phase and the perturbation phase.

The forced event insertion is similar to the *blow-up* operation from [Kos05]. The two main differences are that the suitable timeslots are picked according to an exponentially distributed variable and that there is no attempt to place additional events in p after scheduling e in p. According to our experiments there is no benefit in trying to place additional events in p in order to fill up the rooms.

Combining Construction and Perturbation

Algorithm 5 gives an overview of the algorithm KEMPEINSERTION as proposed in [MW10b*], with a focus on how the construction and the perturbation are combined. The algorithm terminates when a feasible timetable has been found or, depending on the experimental setup, when a timeout or a maximum number of iterations has been reached. The loop in line 1 contains the constructive part of the algorithm, which essentially uses EVENTINSERTION to solve the EIP for events picked uniformly at random (uar) from the list of unscheduled events. The perturbation phase of KEMPEINSERTION starts in line 3, if at most m of the

Algorithm 5: KEMPEINSERTION

input : UCTP instance $\mathcal{I}$, integers k, ℓ, m, exponential distribution parameter λ
in/out : partial timetable τ

$\tau_{best} \longleftarrow \tau$
$\mu \longleftarrow$ list of unscheduled events
while *stopping criterion not met*:
  /* construction phase */
1 **for** $\min\{k, |\mu|\}$ *iterations*:
    pick event e from μ uar
    foreach *timeslot p available for e*:
      solve EIP ($\mathcal{I}, \tau, e, p$) using EVENTINSERTION
      apply Kempe-exchanges found by EVENTINSERTION to τ
      if EVENTINSERTION *indicated success*:
        schedule e in p, remove e from μ if successful
        update τ_{best} if necessary
        break
2   **else:** gather conflicts of e in p
  /* perturcation phase */
3 **if** EVENTINSERTION *was successful at most m times*:
    for *ℓ iterations*:
      pick event e from μ uar
      pick available timeslot p for e (involves λ, see text)
      remove all events conflicting with e in p from τ
      schedule e in p
      update μ
return τ_{best}

$\max\{k, |\mu|\}$ attempts to solving the EIP has been successful in the construction phase, where k is an algorithm parameter and μ is a list of unscheduled events.

Clearly, finding a good balance between construction and perturbation is crucial. The parameter k determines how many attempts to inserting events in the timetable are performed before considering the perturbation. Furthermore, the parameter m can be used to put more focus on the perturbation, that is, increasing m makes the perturbation more likely. The two parameters ℓ and λ set the strength of the perturbation. The parameter ℓ sets the number of perturbation operations that are performed. If ℓ is too small, then KEMPEINSERTION may get stuck in local optima and if ℓ is too large then too many events are removed from the timetable, which hinders progress. During the perturbation, the exponential distribution parameter λ determines how many events are likely to be removed from the timetable. In some sense, it determines the strength of a single perturbation operation. In the original work [MW10b*], λ and ℓ were set to 1.0 and 1, respectively. In Sect. 2.5 we will present the results of an automatic tuning of the parameter choices and the resulting impact on the performance of KEMPEINSERTION.

2.4.3 SAT Encoding of the UCTP

In order to motivate the use of certain specialized heuristics for a given problem domain, it is beneficial to compare their performance to that of standard tools such as SAT solvers and Integer Linear Programming (ILP) solvers. In this section, we will provide a formalization of the UCTP problem in terms of a Boolean formula so we can compare the performance of the KIH and other heuristics for the UCTP to that of a SAT solver. Generally, SAT solvers are very good at solving search problems.

We will first briefly introduce the necessary formalisms and then provide a reduction from the UCTP to the SAT problem.

Boolean Satisfiability

The Boolean satisfiability (SAT) problem is a classical NP-complete problem, see e.g. [GJ79, Section 2.6]. Given a Boolean formula φ in Conjunctive Normal Form (CNF), the task is to decide if there is an assignment of the variables to the truth values $\{\texttt{TRUE}, \texttt{FALSE}\}$ that satisfies φ.

Boolean Formulas Let $\mathcal{A}$ be a finite set of propositional variables. A *literal* is either a variable $x \in \mathcal{A}$ or its negation $\neg x$. A *clause* is a disjunction ($\vee$) of literals. A Boolean formula in Conjunctive Normal Form (CNF) is a conjunction ($\wedge$) of clauses. An *assignment* of the variables is a mapping $a : \mathcal{A} \to \{\texttt{TRUE}, \texttt{FALSE}\}$. By applying the usual semantics of conjunction, disjunction and negation, an assignment determines the truth value of a CNF formula φ over $\mathcal{A}$: A literal $\ell = x$ is true iff $a(x) = \texttt{TRUE}$ and $\ell = \neg x$ is true iff $a(x) = \texttt{FALSE}$. A clause is true, iff at least one of its literals is true and a CNF formula is true iff each clause is true. For convenience, we will denote by $a(\varphi)$ the truth value of φ resulting from the assignment a of the variables. The formula φ is satisfiable if there is an assignment a such that $a(\varphi) = \texttt{TRUE}$. For any Boolean formula φ constructed from the variables $\mathcal{A}$, parentheses and the logical connectives $\{\wedge, \vee, \neg\}$ a formula φ' in CNF can be given that is logically equivalent. That is, any assignment that satisfies φ also satisfies φ' and vice versa. However, the transformation of φ into φ' may result in an exponential blowup of the number of connectives. We represent a CNF formula as a set of its clauses, so we can create a CNF $\varphi \wedge \varphi'$ from two CNF formulas φ and φ' over $\mathcal{A}$ by taking the set union $\varphi \cup \varphi'$. An assignment satisfies $\varphi \wedge \varphi'$ if and only if it satisfies both φ and φ'.

Unit Propagation A *unit clause* is a clause that contains a single literal. An inference rule used by many SAT solvers is *unit clause propagation*, or *unit propagation* for short. In particular, unit propagation is the main inference rule of all SAT solvers based on the Davis–Putnam–Logemann–Loveland (DPLL) procedure [DLL62], including, for example, MiniSAT [ES04] and (z)Chaff [MMZ+01]. Let φ be a CNF formula with a unit clause that contains the literal $\ell \in \{x, \neg x\}$, $x \in \mathcal{A}$.

Then the value $a(x)$ is the same for any assignment a that satisfies φ: If $\ell = x$ then $a(x) = \texttt{TRUE}$ since otherwise $a(\varphi) = \texttt{FALSE}$. Similarly, if $\ell = \neg x$ then necessarily $a(x) = \texttt{FALSE}$. Now, all clauses that contain literals in $\{x, \neg x\}$ can be simplified by substituting the value of x. This process can be performed repeatedly if new unit clauses appear due to the substitution. The advantage of unit propagation is a SAT solver can use it to efficiently prune the search space.

Cardinality Constraints It is a common requirement in SAT encodings of combinatorial problems that at least, at most, or exactly k variables from a given set $X \subseteq \mathcal{A}$ are true. Such requirements are called *cardinality constraints*. The main challenge is to find economical encodings of these constraints with respect to the number of clauses and auxiliary variables required. For our purpose of modeling the UCTP we need to formulate the requirement that among n variables $x_1, \ldots, x_n \in \mathcal{A}$ exactly one is true, which is denoted by $=_1(x_1, \ldots, x_n)$. We will express this constraint by a conjunction of the constraints $\leq_1(x_1, \ldots, x_n)$ (at most one of the variables is true) and $\geq_1(x_1, \ldots, x_n)$ (at least one of the variables is true). The latter can be realized in a straight-forward fashion by the following clause:

$$\geq_1(x_1, \ldots, x_n) = (x_1 \vee \ldots \vee x_n) \enspace .$$

The $\leq_1$ constraint can be modeled by prohibiting for any choice of two variables that both are true:

$$\leq_1(x_1, \ldots, x_n) = \bigwedge_{\substack{i,j \in \{1,\ldots,n\} \\ i>j}} (\neg x_i \vee \neg x_j) \enspace .$$

This CNF encoding of $\leq_1$, referred to as the pairwise encoding, can be improved by various techniques that have been developed for the $\leq_k$ constraint, which is true iff at most k of the given variables are true [Sin05, AANORC09, FG10, Che10]. However, for the SAT encoding of the UCTP we use the pairwise encoding given above with good results.

Reduction of UCTP to SAT

There are many different ways to model the UCTP as a CNF. The differences are manifest mainly in the number of variables, the number of clauses and the lengths of the clauses required to describe the problem, and certainly how well a SAT solver can handle the resulting instances. Depending on the problem encoding, SAT solvers may show quite different performance. See for example [VG08] for an investigation of the impact of different encodings of the graph coloring problem on the performance of various SAT solvers. Our objective is to give a problem encoding of the UCTP that is succinct and enables the SAT solver to make good use of unit propagation. The succinctness is mainly important to ensure that the

SAT instances generated from large UCTP instances fit in the main memory on commodity hardware, which can be a problem with other encodings.

Let $\mathcal{I}$ be a UCTP instance with room and timeslot availability requirements. In the following, let $e_1, \ldots, e_{|E|}$ be the events, $r_1, \ldots, r_{|R|}$ be the rooms, and $p_1, \ldots, p_{|P|}$ be the timeslots given by the instance. The choice of the ordering is irrelevant since any two different orderings will just result in different variable names. We give a CNF $\varphi(\mathcal{I})$ such that $\varphi(\mathcal{I})$ is satisfiable if and only of $\mathcal{I}$ has a feasible solution.

The SAT encoding $\varphi(\mathcal{I})$ has the following variables:

- event to room mapping: $\{er_{e,r} \mid e \in E, r \in R\}$

$$er_{e,r} = \begin{cases} \texttt{TRUE} & \text{if } e \text{ is scheduled in room } r \\ \texttt{FALSE} & \text{otherwise} \end{cases},$$

- event to timeslot mapping: $\{ep_{e,p} \mid e \in E, p \in P\}$

$$ep_{e,p} = \begin{cases} \texttt{TRUE} & \text{if } e \text{ is scheduled in timeslot } p \\ \texttt{FALSE} & \text{otherwise} \end{cases},$$

- event conflicts: $\{conf_{e_i,e_j} \mid i, j \in \{1, \ldots, |E|\}, i > j\}$

$$conf_{e_i,e_j} = \begin{cases} \texttt{TRUE} & \text{if } e_i \text{ and } e_j \text{ are conflicting} \\ \texttt{FALSE} & \text{otherwise} \end{cases},$$

- room availability: $\{ra_{e,r} \mid e \in E, r \in R\}$

$$ra_{e,r} = \begin{cases} \texttt{TRUE} & \text{if } e \text{ can be scheduled in room } r \\ \texttt{FALSE} & \text{otherwise} \end{cases},$$

- timeslot availability: $\{pa_{e,p} \mid e \in E, p \in P\}$

$$pa_{e,p} = \begin{cases} \texttt{TRUE} & \text{if } e \text{ can be scheduled in timeslot } p \\ \texttt{FALSE} & \text{otherwise} \end{cases},$$

We also introduce the following auxiliary variables.

- "same room"-indicators $\{sr_{e_i,e_j} \mid i, j \in \{1, \ldots, |E|\}, i > j\}$
- "same timeslot"-indicators $\{sp_{e_i,e_j} \mid i, j \in \{1, \ldots, |E|\}, i > j\}$

All relations between two events, i. e., the event conflicts and the "same room" and "same timeslot" properties, are symmetric. The ordering of the events lets us exploit this symmetry and reduce the number of variables required to encode these relations. The total number of variables in $\varphi(\mathcal{I})$ is given by:

$$\mathrm{var}(\varphi(\mathcal{I})) = 2|E| \cdot (|R| + |P|) + 3 \cdot \binom{|E|}{2} . \tag{2.4}$$

In the following, we will characterize the UCTP by a CNF formula. First of all, we make sure that the auxiliary variables have the following properties: A "same room"-indicator variable should be set to TRUE whenever the two corresponding events are scheduled in the same room, irrespective of their timeslot assignments. That is, for a timeslot $p \in P$ and any two events $e_i, e_j, i > j$,

$$ep_{e_i,p} \wedge ep_{e_j,p} \longrightarrow sp_{e_i,e_j} .$$

This property is realized by the following clauses:

$$\bigwedge_{\substack{r \in R \\ i,j \in \{1,\dots,|E|\} \\ i>j}} \left(\neg er_{e_i,r} \vee \neg er_{e_j,r} \vee sr_{e_i,e_j}\right) . \tag{2.5}$$

Similarly, a "same timeslot"-indicators variable should be TRUE if the two corresponding events are scheduled in the same timeslot. This is realized by the following clauses:

$$\bigwedge_{\substack{p \in P \\ i,j \in \{1,\dots,|E|\} \\ i>j}} \left(\neg ep_{e_i,p} \vee \neg ep_{e_j,p} \vee sp_{e_i,e_j}\right) . \tag{2.6}$$

Now we can capture all properties of the UCTP in a straight forward fashion:

- Each event is scheduled in exactly one room:

$$\bigwedge_{e \in E} \left(\leq_1(er_{e,r_1}, \dots, er_{e,r_{|R|}}) \wedge \geq_1(er_{e,r_1}, \dots, er_{e,r_{|R|}})\right) \tag{2.7}$$

- Each event is scheduled in exactly one timeslot:

$$\bigwedge_{e \in E} \left(\leq_1(ep_{e,p_1}, \dots, ep_{e,p_{|R|}}) \wedge \geq_1(ep_{e,p_1}, \dots, ep_{e,p_{|R|}})\right) \tag{2.8}$$

- No two events are in the same room and in the same timeslot:

$$\bigwedge_{\substack{r \in R \\ i,j \in \{1,\dots,|E|\} \\ i>j}} \left(\neg sr_{e_i,e_j} \vee \neg ep_{e_i,e_j}\right) \tag{2.9}$$

- No two conflicting events are in the same timeslot:

$$\bigwedge_{\substack{r\in R\\ i,j\in\{1,\dots,|E|\}\\ i>j}} \left(\neg sp_{e_i,e_j} \vee \neg conf_{e_i,e_j}\right) \tag{2.10}$$

- Room availability requirements are respected:

$$\bigwedge_{\substack{r\in R\\ e\in E}} (ra_{e,r} \vee \neg er_{e,r}) \tag{2.11}$$

- Period availability requirements are respected:

$$\bigwedge_{\substack{p\in P\\ e\in E}} (pa_{e,p} \vee \neg ep_{e,p}) \tag{2.12}$$

The following unit clauses encode the room and timeslot availability for each event, as well as the event conflicts:

$$\bigwedge_{\substack{p\in P\\ e\in E}} \begin{cases} pa_{e,p} & \text{if } p \text{ is available for } e\\ \neg pa_{e,p} & \text{otherwise} \end{cases}, \tag{2.13}$$

$$\bigwedge_{r\in R,\, e\in E} \begin{cases} ra_{e,r} & \text{if } r \text{ is available for } e\\ \neg ra_{e,r} & \text{otherwise} \end{cases}, \tag{2.14}$$

$$\bigwedge_{\substack{i,j\in\{1,\dots,|E|\}\\ i>j}} \begin{cases} conf_{e_i,e_j} & \text{if } e_i \text{ is conflicting with } e_j\\ \neg conf_{e_i,e_j} & \text{otherwise} \end{cases}, \tag{2.15}$$

Combining the CNF formulas (2.5)–(2.15) into a single CNF formula (technically by taking the set union) yields the CNF formula $\varphi(\mathcal{I})$. All constraints need to be satisfied simultaneously. It is evident from the construction that $\mathcal{I}$ admits a feasible timetable if and only if $\varphi(\mathcal{I})$ is satisfiable. We will briefly discuss a subtlety in the role of the auxiliary variables. Note that in Eqs. (2.6) and (2.5) the auxiliary variables are necessarily true whenever two events are in the same timeslot and room, respectively. If the events are not in the same timeslot or room, the auxiliary variables may still be set to true. This however, adds restrictions on the problem, see Eqs. (2.9) and (2.10). Therefore, no matter how the values of the auxiliary variables are chosen in these cases, a feasible assignment of the variables $er_{e,r}$ and $ep_{e,p}$, $e \in E, r \in R, p \in P$ encodes an assignment of the events to resources that satisfies all constraints, i.e. a feasible timetable.

The number of clauses of $\varphi(\mathcal{I})$ is given by:

$$\text{size}(\varphi(\mathcal{I})) = 2 + 2 \cdot |E| \cdot (|P| + |R|) + (1 + 3 \cdot |R| + |P|) \cdot \binom{|E|}{2} + |E| \cdot \left(\binom{|R|}{2} + \binom{|P|}{2} \right) . \tag{2.16}$$

Since we can assume that $|E| = |R| = O(|\mathcal{I}|)$ we have $\text{size}(\varphi(\mathcal{I})) = O(|\mathcal{I}|^3)$. The proposed CNF encoding of the UCTP is very unit propagation friendly: Eqs. (2.13)–(2.15) can be eliminated immediately and all other clauses except the $2 \cdot |E|$ $\geq_1$-constraints have length at most three, so there is potential for eliminating more clauses quickly.

2.5 Experimental Evaluation of the Kempe Insertion Heuristic

The overall goal of this section is a comparison of the performance of the KIH against other approaches from the literature that focus on finding feasible timetables. Before this comparison, we will investigate the following two questions:

1. What is the impact of the preprocessing phase in the KIH?
2. What is the performance of the KIH compared to other approaches from the literature on suitable benchmark instances?

Most state-of-the-art approaches that deal with finding feasible timetables use some form of preprocessing, see for example [TBM07,MW10b*,LZC11,CDGS12]. The intention is to quickly create an initial timetable, which is used as a starting point for a more elaborate algorithm that deals with the remaining hard constraint violations. The first question is motivated by the fact that despite this widely used decomposition into a preprocessing and an optimization phase, there is no rigorous investigation of the quality of the timetables produced by various preprocessing algorithms. That is, it is unclear with how many constraint violations the elaborate optimization approaches have to deal for different preprocessing algorithms. We will provide experimental data on the distance to feasibility of the timetables produced by five different preprocessing heuristics, including the one used in [TBM07, MW10b*].

To investigate the second question, we first perform an automated tuning of the KIH parameter settings. Note that the performance of any of the popular meta-heuristic approaches depends on the parameter settings. For example, the performance of SA depends on choosing good temperature levels for the problem at hand [KGV83]. Similarly, EA performance naturally depends on mutation and crossover rates. The KIH has a number of parameters and for our investigation of the second question we are interested in tuning these parameters in a systematic fashion to a set of UCTP benchmark instances. In order to determine good parameter settings, we first check if there are significant differences between a number of tuned parameter settings. We include in this comparison the parameter choice from the original work [MW10b*], which has been determined by trial and error without using statistical methods such as hypothesis testing.

We will rely on the parameter settings determined by the systematic tuning in our overall comparison of the KIH against other approaches from the literature. As noted by [TBM07], there are only few approaches that focus on finding feasible timetables. One reason for this is certainly, that on many popular benchmark instances, for example, the instances from the timetabling competition 2002 [PGRD], it is rather easy to find feasible timetables [LP07] and the actual challenge is to minimize the soft constraint violations. As a remedy, Lewis and Paechter introduced three sets of artificially generated UCTP instances, referred to as small, medium and big, with a focus on finding feasible timetables rather than soft constraint satisfaction [LP07, LPb]. The small instances have between 200 and 225 events and 5 or 6 rooms, the medium instances have between 390 and 425 events and 10 or 11 rooms, and the big instances have between 1,000 and 1,075 events and 25 to 28 rooms. The instances sets were created by selecting from a set of randomly generated instances 20 for each set on which the algorithms from [AL05, LP04] performed poorly [LP07]. These instances are indeed challenging, and as of today, none of the approaches from the literature is able to produce feasible timetables on all instances. We perform our two experiments on these instances. In our overall comparison of approaches for creating feasible timetables, we will focus on approaches that have been evaluated on the Lewis and Paechter instance sets.

Preliminaries: Algorithm Configuration

The goal of automated algorithm tuning is to select parameters of a given algorithm such that it performs best on a given set of training instances. This problem has been formalized in terms of the Algorithm Configuration Problem (ACP) [BYBS10]. A choice of values for the parameters of an algorithm is referred to as *configuration*. The task is to find a configuration such that the expected value of a cost measure—in our setting, for example, the distance to feasibility—is minimal over a set of benchmark instances. We use the I-RACE algorithm, from [BYBS10], to perform the tuning task. An implementation of this algorithm is available as a package for the R statistics software [LIDLSB11, R D08] and a brief outline of the approach is given below. We will use statistical hypothesis testing in order to determine if there is a significant benefit from tuning the algorithm parameters in Experiment 2.

The Iterated F-RACE (I-RACE) Algorithm [BYBS10] F-RACE evaluates a given set of candidate configurations on a set of benchmark instances in order to determine the best configurations according to some quality measure. For minimization problems, the measure is typically the expected cost generated by a configuration. The naïve approach is to just evaluate all given configurations on all instances sufficiently often. The motivation of F-RACE is to make better use of computational resources by discarding configurations as soon as there is enough statistical evidence indicating that they perform worse than the other configurations. Hence, if a number of configurations can be discarded, more time can be spent on evaluating the performance of the well-performing configurations. In each step of the algorithm, the configurations which have not been discarded yet are evaluated on a particular instance. The null hypothesis is that all remaining configurations show equal performance. If the null hypothesis is rejected according to Friedman's two-way analysis of variance by ranks [Fri37, Con06] (i. e., the Friedman test) with a significance level $\alpha = 0.05$, then appropriate post-hoc tests are performed in order to decide which configurations are discarded. If there is not enough statistical evidence to reject the null hypothesis, all configurations are kept for the next iteration. While F-RACE evaluates a fixed set of configurations, I-RACE iteratively executes F-RACE and uses the results in order to bias the sampling of new configurations and thus offers arguably better exploration of the configuration space.

2.5.1 Experiment 1: The Impact of Preprocessing

Algorithms for the UCTP typically use some computationally cheap constructive preprocessing step that generates an initial timetable used as a starting point for more elaborate techniques that deal with the remaining hard constraint violations. Examples of such approaches can be found in [Kos05, TBM07, MW10b*, LZC11, CDGS12]. The motivation for this decomposition is to save computation for the sophisticated techniques. To the best of our knowledge, no detailed data has been published on the constraint violations that remain after the preprocessing phase on the Lewis and Paechter instances from [LPb]. Thus, it is not clear, how much work is left to be done by the elaborate techniques using different preprocessing approaches. Here, we will provide empirical data on the distance to feasibility that can be attained by five different sequential heuristics including the one used in [TBM07, MW10b*].

Experiment Design The behavior of a sequential heuristic is determined by an event selection policy and a resource selection policy. See Sect. 2.4.1 for a list of policies from the literature. We denote by A/B the sequential heuristic that uses event selection policy A and resource selection policy B. Our comparison includes the heuristics LSD/RO, LSDLD/RO, LD/SO, LSD/SO and LSDLD/SO. The two resource selection policies RO and SO were discussed in Sects. 2.4.1 and 2.4.2, respectively. A run of a sequential heuristic is evaluated with respect to the distance to feasibility of the resulting timetable. We do not consider the runtimes of the sequential heuristics, because the differences are negligible due to the overall very similar algorithmic scheme. We collect statistical information over 500 independent runs per heuristic and per instance, namely the attained minimum, average, and maximum distance to feasibility. Our intention for this is twofold: First, we want to get a sufficient sample size to allow for conclusions about the expected distance to feasibility for each algorithm and instance, and second, we intend to independently verify the following observation by Tuga et al. in [TBM07]: According to their findings, running the LSDLD/RO heuristic 500 times per instance produces feasible timetables on 13 out of the 60 benchmark instances.

Data Table 2.3 shows for each of the 60 benchmark instances from [LPb] the minimum, average, and maximum number of events left unscheduled by the five different sequential heuristics. The last row of Table 2.3 summarizes for each heuristic how often it performed at least as good as any of the other heuristics with respect to the minimum/average/maximum number of events that were left unscheduled.

Results We can conclude from the data shown in Table 2.3 that there is quite some variation in the distance to feasibility, even for different runs of the same heuristic. Thus, it is reasonable to run the sequential heuristics multiple times in the preprocessing phase in order to reduce the variance of the minimum distance to feasibility attained. According to Tuga et al. [TBM07], feasible solutions to 13 out of 60 instance can be found by running the LSDLD/RO heuristic 500 times per instance. We were able to confirm this result: According to the data in Table 2.3, feasible timetables were found on 11 instances by this sequential heuristic and on several other instances the resulting timetables were close to feasibility. Overall, the SO resource selection policy seems to generate the best results in combination with the LSD event selection policy, with or without using LD for breaking ties.

2.5.2 Experiment 2: Algorithm Configuration and Evaluation

The KIH proposed in [MW10b*] has a number of parameters that have some influence on its behavior, for example, the balance between construction and perturbation. Naturally, we are interested in picking a configuration such that the algorithm performs best on the instances of interest. In [MW10b*], this task has been performed by hand with satisfactory results. In order to check in how far the performance of the algorithm can be improved by altering the configuration and how much the quality of the results depends on the configuration, we perform an automated parameter tuning of the KIH using the I-RACE algorithm (see p. 58).

Fixed and Varied Parameters

Recall from Sect. 2.4.2, that the algorithm KEMPEINSERTION has parameters k, ℓ, m, and λ. The parameters k and m determine the relative importance of the construction phase and the perturbation phase, and ℓ is the perturbation strength.

Table 2.3 Results of Experiment 1: minimum/average/maximum distance to feasibility attained by five different sequential heuristics on the Lewis and Paechter instances from [LPb]

	LSD/RO	LSDLD/RO	LD/SO	LSD/SO	LSDLD/SO
small_1	**0**/0.4/4	**0**/0.4/**3**	**0**/**0.3**/4	**0**/7.4/15	4/6.7/8
small_2	**0/0.0/0**	**0/0.0/0**	**0/0.0/0**	**0/0.0/0**	**0/0.0/0**
small_3	6/13.3/23	7/13.6/23	10/21.7/33	**0**/2.2/13	1/**1.0/1**
small_4	**0**/5.8/11	**0**/6.0/13	**0**/5.1/12	1/6.9/19	**0/3.1/8**
small_5	10/17.8/28	10/17.0/24	14/22.2/30	**1/8.3**/17	7/8.7/**12**
small_6	**0/0.0/0**	**0/0.0/0**	**0/0.0/0**	**0**/2.4/7	1/3.2/6
small_7	3/7.6/14	2/6.9/14	2/7.9/17	**0**/0.5/7	**0/0.0/0**
small_8	23/32.1/42	23/31.6/39	23/34.2/44	**16/25.5**/36	23/26.9/**30**
small_9	32/52.1/69	44/59.2/72	35/49.6/68	**8**/23.4/43	18/**23.0/28**
small_10	14/28.9/57	10/18.1/23	34/45.0/59	**0**/7.3/32	**0/0.2/2**
small_11	**0**/0.0/4	**0/0.0/2**	**0**/0.4/5	**0**/0.8/4	**0**/1.6/5
small_12	**0/0.2/4**	**0**/0.6/5	2/5.5/10	3/10.6/15	6/10.7/14
small_13	26/48.7/75	33/45.4/53	30/44.4/62	**2**/22.0/58	7/**12.1/14**
small_14	31/39.5/52	28/41.0/52	29/41.3/54	**16/27.7**/42	20/30.5/**35**
small_15	**0**/5.7/14	2/7.1/17	**0**/4.3/13	**0/0.0/0**	**0/0.0/0**
small_16	12/22.7/39	12/21.8/43	14/28.5/40	**0**/2.3/16	**0/0.0/0**
small_17	31/42.3/53	26/44.9/53	25/38.1/51	**0**/9.8/29	3/**3.6/5**
small_18	9/19.5/30	14/22.4/29	**4/8.4/15**	12/22.8/30	14/18.4/23
small_19	47/60.4/75	47/60.4/75	37/50.8/64	**1/21.6**/55	24/27.0/**30**
small_20	**0**/0.4/5	**0/0.4**/5	**0**/1.1/**4**	1/4.4/14	**0**/2.5/5
med_1	**0**/5.8/15	**0/5.6/12**	1/7.6/16	5/12.7/25	4/11.4/18
med_2	**0**/5.7/18	**0/4.6/12**	1/9.4/19	4/15.3/29	6/10.4/17
med_3	**1/10.0/21**	4/12.9/22	11/21.1/33	8/17.8/30	19/21.1/25
med_4	**5/13.0**/24	7/15.1/24	13/21.1/31	12/23.9/40	6/13.9/**22**
med_5	22/38.2/49	26/38.8/49	22/36.6/47	**19**/29.2/44	20/**26.2/33**
med_6	29/47.7/61	38/50.5/67	30/48.1/63	**15**/31.8/43	28/**30.9/33**
med_7	57/71.7/90	62/74.8/91	81/99.5/114	**44**/59.3/76	49/**58.1/68**
med_8	49/61.9/77	48/63.5/77	50/65.5/79	**23**/36.3/52	33/**34.9/39**
med_9	52/73.6/96	65/79.5/97	73/90.6/110	**40/56.8**/81	59/63.2/**69**
med_10	**0**/0.4/5	**0/0.4/4**	**0**/1.1/6	1/11.8/29	5/10.2/18
med_11	27/41.6/56	28/40.6/56	28/41.2/54	**0**/10.3/37	1/**5.2/12**
med_12	3/10.8/19	3/10.5/19	5/16.1/33	**0/3.0/7**	2/4.3/**7**
med_13	47/63.2/78	53/65.5/78	58/71.9/87	**13/26.1**/41	32/34.4/**38**
med_14	16/28.8/44	15/29.6/42	20/32.8/47	**0**/5.7/23	2/**3.0/4**
med_15	21/30.9/42	18/30.2/39	24/34.5/44	**6**/22.9/46	8/**14.4/23**
med_16	97/116.9/133	99/117.3/137	99/120.5/145	**27/62.4**/104	63/65.6/**68**
med_17	58/75.0/91	60/74.7/88	48/70.4/95	**0**/20.3/52	13/**14.5/24**
med_18	126/147.6/168	126/150.2/173	112/137.0/164	**21/79.3**/131	80/86.8/**94**
med_19	114/133.4/155	110/132.3/150	109/135.6/154	**31/79.9**/118	80/80.2/**81**
med_20	78/93.0/113	78/94.3/113	83/103.1/121	**1**/43.0/79	33/**33.0/33**

(continued)

Table 2.3 (continued)

	LSD/RO	LSDLD/RO	LD/SO	LSD/SO	LSDLD/SO
big_1	**0**/1.8/6	**0**/1.4/8	**0/0.5/4**	2/16.7/39	8/18.1/30
big_2	**9/21.9/36**	**9**/26.0/40	28/42.2/60	34/49.0/67	35/47.8/61
big_3	**14/28.4/41**	19/34.2/46	21/32.7/46	31/45.4/60	31/42.4/55
big_4	44/**58.7**/84	**39**/58.8/**72**	68/88.3/107	55/79.2/111	57/67.7/77
big_5	**53/74.8/89**	65/78.2/93	93/111.8/129	78/92.9/110	80/93.2/107
big_6	95/117.4/137	98/115.8/137	158/182.6/209	90/113.5/138	**88/101.9/119**
big_7	170/196.0/230	183/209.4/237	256/279.5/306	**159/187.5**/223	181/192.4/**212**
big_8	**52/70.3**/90	53/70.5/**89**	87/109.3/126	59/75.5/93	63/79.7/93
big_9	**39**/55.6/76	41/55.5/71	66/82.4/99	44/62.8/88	43/**54.0/66**
big_10	**60**/75.4/93	64/**75.0/89**	101/117.7/137	68/93.2/129	68/79.6/93
big_11	**55/71.4/85**	57/72.9/91	100/117.4/137	74/94.2/114	70/83.3/96
big_12	41/59.1/80	47/62.4/78	29/42.1/59	28/44.3/66	**27/38.5/51**
big_13	52/73.5/90	60/71.3/84	51/73.6/89	**41**/59.9/78	43/**52.1/60**
big_14	47/73.9/93	58/74.8/91	37/56.3/71	**36**/52.7/79	38/**50.3/68**
big_15	146/178.5/209	155/182.5/203	214/242.1/275	**97/128.4/165**	148/152.8/166
big_16	133/157.2/180	136/158.9/180	127/155.4/184	36/64.9/96	**27/38.2/47**
big_17	339/364.5/394	341/368.9/393	345/388.3/427	**189/257.2**/318	254/272.3/**293**
big_18	221/246.0/275	225/254.5/285	255/293.2/329	**107/153.0**/217	154/163.8/**177**
big_19	284/306.6/334	287/313.4/333	331/353.7/378	**194/239.3**/298	228/245.3/**257**
big_20	233/269.4/294	249/279.2/310	248/287.1/331	**120**/164.5/211	136/**160.2/174**
$\sum$	21/11/8	13/8/10	10/5/5	**39**/17/4	11/**26/41**

Entries in bold face mark the best performing heuristic in each category

The parameter λ is the decay rate of an exponentially distributed random variable that is used in order to select which event is selected to be inserted in the timetable in the perturbation phase. The unscheduled events are arranged in non-decreasing order by the number of conflicts in a certain timeslot and an event is selected to be scheduled in this timeslot according to the exponential distribution. We vary the four parameters k, ℓ, m, and λ in the tuning process. We also vary the event selection policy P_1 and resource selection policy P_2 of the preprocessing in order to check, which of the sequential heuristics performs best in combination with KEMPEINSERTION. For P_1, we consider the LSD and the LSDLD policies, and for P_2 the RO and SO policies.

We concluded from Experiment 1 that sequential heuristics should be run several times in the preprocessing phase in order to compensate for that rather high variance of the distance to feasibility. In the current experiment we set the number of trial runs of the sequential heuristics to 15, which is considerably lower than what Tuga et al. used in [TBM07]. To give some justification for this choice, we iteratively performed the following experiment: We measured the difference between the minimum distances to feasibility attained by 15 and 500 runs of a sequential heuristic in percent relative to the total number of events. Table 2.4 shows the average outcome and the standard deviation over 50 runs per instance for each

Table 2.4 Mean and standard deviation of the difference in the distance to feasibility between running the sequential heuristics 15 and 500 times in percent relative to the total number of events

Instance set	LSD/RO	LSDLD/RO	LSD/SO	LSDLD/SO
Small	1.36 ± 1.50	1.39 ± 1.50	0.35 ± 0.62	0.92 ± 1.22
Med	0.60 ± 0.78	1.40 ± 0.99	0.20 ± 0.31	1.80 ± 2.30
Big	0.74 ± 0.51	0.84 ± 0.56	0.40 ± 0.33	1.00 ± 0.85

instance set and each sequential heuristic. We consider the differences between 15 and 500 runs to be sufficiently small to cut down the number of iterations of the sequential heuristic to 15.

Performance Measures

We measure the performance of a run of the KIH (preprocessing followed by KEMPEINSERTION) according to two criteria, the runtime and the distance to feasibility. For a comparison of different configurations of the KIH we combine these two criteria in the following way: Let (t_1, d_1) and (t_2, d_2) be the measured performance of two runs r_1 and r_2, where t_1 and t_2 are the runtimes given in seconds and d_1 and d_2 are the distances to feasibility. Then r_1 showed better performance than r_2 if and only if $(t_1, d_1) \prec_{lex} (t_2, d_2)$. Intuitively, a run that produces a feasible timetable is always better than one producing an infeasible timetable. Among two runs that produce feasible timetables, one that takes less runtime is to be preferred and among two runs that produce infeasible timetables, one that leaves fewer events unscheduled is to be preferred. The performance of the KIH is compared to the performance of various algorithms from the literature according to the minimum and average number distance to feasibility over a number of runs as well as the average runtime used.

Experiment Design

We will first discuss the experiment designs that were used for evaluating different approaches from the literature in order to motivate the design of our experiment. Lewis and Paechter evaluate their GGA algorithms using a timeout of 30, 200, and 800 s for the small, medium and big instance sets, respectively [LP07]. Liu et al. evaluate their CB algorithm using the same timeouts. However, also set a fixed maximum number of iterations and on some instances their algorithm apparently terminates before the timeout, even if no feasible timetable has been found [LZC11]. Tuga et al. use timeouts of 200, 400, and 1,000 s for the small, medium and big instance sets, respectively [TBM07]. Ceschia et al. use a fixed number of 10^8 iterations of their SA-based approach and no timeout [CDGS12]. Due to the different experiment design, it is quite a challenge to make a fair comparison between the different approaches.

There are also some more fundamental issues with the different experiment designs that require some consideration, most of which have been discussed in [Joh02]. Using a timeout can be quite misleading because of ever-increasing CPU clock speeds and architectural improvements of the hardware. Certainly, 3 s of CPU time in 2013 is different from 30 s of CPU time in 2007, or in 1990, which impairs reproducibility in particular of older results. Additionally, in the presence of a timeout the efficiency of the implementation has a significant influence on the results. So, in this experiment design, there is no satisfactory answer to the question whether an improved solution quality should be attributed to a proposed algorithm, the implementation or the underlying hardware. Setting a maximum number of iterations or evaluations of the cost function avoids these pitfalls but introduces a new one: different heuristic approaches may perform a vastly different amount of work per iteration. For example, performing a single iteration of KEMPEINSERTION is quite costly in terms of CPU time, in particular when the EIP cannot be solved, because a number of Kempe-exchanges and maximum cardinality matchings are computed. In comparison, performing an iteration of SA with a simple neighborhood structure is much quicker. Furthermore, a comparison between heuristics and exact approaches such as SAT or ILP solvers based on a maximum number of iterations is non-sensical. In conclusion, neither a timeout nor a fixed iteration count is the silver bullet for conducting comparable computer experiments.

For the parameter tuning and the overall comparison of KIH with the other approaches we will use the timeouts suggested by Lewis and Paechter [LP07] in our experiments, which allows for a somewhat fair comparison of our results and the results in [LP07, TBM07, LZC11], and to some extent the ones in [CDGS12]. Furthermore, we will compare the results of these heuristics to those produced by a SAT solver using the formulation from Sect. 2.4.3. All experiments were performed on a single core of a 3.0 GHz i7 machine. Certainly, in our comparison of the heuristic approaches, we cannot cure the flaws that are inherent in using timeouts, however, this machine should show comparable performance with the machines used in [LP07, TBM07, LZC11].

Following [CDGS12], we perform the parameter tuning separately on each set of instances from [LPb]. For each run of the KIH with a given configuration, we use the timeouts suggested by [LP07], that is, 30 s for the small instances, 200 s for the medium instances and 800 s for the big instances. After some preliminary screening, the parameter domains used by I-RACE were set to the values shown in Table 2.5.

In total, the total computational budget of I-RACE was set to 1,000 runs of the KIH on each instance set. The instances that present no challenge to the KIH were removed from the instance sets. Thus, the tuning is focused on getting good performance on the challenging instances, which offer more potential for improvements. The assumption is that a configuration tuned to a restricted instance set also performs well on the remaining instances. We will comment on this issue later when we discuss the results. The instances shown in Table 2.6 were considered "challenging" and were therefore used as training instances for the automated tuning.

Table 2.5 Parameter domains considered in the automated tuning of the KIH

Parameter	Domain
P_1	{LSD,LSDLD}
P_2	{RO,SO}
ℓ	$[10, 60]$
m	$[1, 5]$
k	$[1, 5]$
λ	$[0.5, 2.0]$

Table 2.6 Training instance sets used in the automated tuning process

Instance set	Training instances
Small	8, 9, 12, 13, 14, 18, 19, 20
Medium	7, 8, 9, 12, 18, 19, 20
Big	4, 5, 6, 7, 8, 9, 10, 11, 15, 17, 19, 20

Table 2.7 KIH configurations for the small, medium and big instances

Configuration	P_1	P_2	ℓ	m	k	λ
(a) Configurations for the instances `small_1`, ..., `small_20`						
A_{small}	LSD	SO	32	2	3	1.20
B_{small}	LSD	RO	17	2	1	1.72
C_{small}	LSD	SO	22	2	4	1.30
D_{small}	LSD	SO	15	2	1	1.85
E [MW10b*]	LSDLD	RO	20	1	1	1.00
(b) Configurations for the instances `med_1`, ..., `med_20`						
A_{med}	LSDLD	SO	13	3	2	1.65
B_{med}	LSDLD	SO	16	3	4	1.15
C_{med}	LSDLD	SO	17	3	2	1.63
D_{med}	LSDLD	SO	11	3	2	1.64
E [MW10b*]	LSDLD	RO	20	1	1	1.00
(c) Configurations for the instances `big_1`, ..., `big_20`						
A_{big}	LSDLD	SO	17	2	5	1.53
B_{big}	LSDLD	SO	23	2	5	1.46
C_{big}	LSDLD	SO	18	2	5	1.67
D_{big}	LSDLD	SO	22	2	2	1.68
E [MW10b*]	LSDLD	RO	20	1	1	1.00

Configurations A–D have been determined by I-RACE for each instance set and configuration E is from [MW10b*]

The output of I-RACE is a set of the best four configurations found for each instance set. We perform 20 independent runs for each of these configurations on each instance of the corresponding (complete) instance set. Additionally, we perform 20 independent runs for each of the 60 instances with the KIH configuration given in [MW10b*], referred to as configuration E. This configuration is also shown in Table 2.7. This configuration has been determined by manual experimentation. Next, we will check if the differences in performance between any two configurations are significant according to the one-tailed Wilcoxon-Mann-Whitney (WMW) test [Wil45, Con06]. That is, for any two configurations X and Y, the null-hypothesis being tested is that the KIH does not show a different performance with configuration X compared to Y. The alternative hypothesis is that X produces better results than Y. We perform the test on samples of 20 independent runs for each configuration on each instance of the corresponding set.

Finally, we compare the performance of the KIH to other approaches from the literature and the SAT-solver `minisat` [ES04] with respect to the minimum and average distance to feasibility and, if available, the average processing time used. We use for each instance set the best configuration among the tuned configurations and the original configuration from [MW10b*]. Thus, we provide an update to the experimental results in [MW10b*] in a slightly modified experimental setup. Please note that in this comparison, a lower distance to feasibility on the difficult instances is more desirable than spending less CPU time on the easier instances. This preference is not present in the performance measure, but rather by the selection of instances that are used for the configuration tuning. The SAT instances were created according to the SAT formulation from Sect. 2.4.3. Overall, we include in our comparison the SAT solver and the data from the literature shown in Table 2.8.

These algorithms were summarized briefly in Sects. 2.4.1 and 2.4.2. This comparison gives a complete picture of the relative performance of published approaches to solving the Lewis and Paechter instances.

In addition to the distance to feasibility and the processing time, we measure the average number of room assignments performed by KEMPEINSERTION per second on each instance. Our motivation is to be able to identify potential scaling issues of the algorithm. Since the number of rooms increases between small, medium, and large instances, we expect the number of room assignments performed per unit time to be significantly lower on the big instances compared to the small and medium instances.

Table 2.8 Algorithms from the literature which are compared to the KIH, along with the corresponding publications and a rough classification of the used techniques

Identifier	Publication	Technique
GGA	Lewis and Paechter [LP07, LPa]	Grouping Genetic Algorithm
LS	Lewis and Paechter [LP07, LPa]	Local Search
KIH	M. and Wanka [MW10b*] (see text)	Kempe Insertion Heuristic
HSA	Tuga et al. [TBM07]	Simulated Annealing
CB	Liu et al. [LZC11]	Clique-based Construction
TSA	Ceschia et al. [CDGS12]	Simulated Annealing

Data

Table 2.7 shows the four best KIH configurations determined by I-RACE for each instance set along with the configuration E from [MW10b*]. The configurations A_{small}–D_{small}, A_{med}–D_{med}, and A_{big}–D_{big} are the best configurations found by I-RACE for the small, medium, and big instances, respectively. Table 2.9 shows the results of the WMW-tests performed to detect performance differences between the configurations. For each pair (X, Y) of configurations, the corresponding cell shows the number instances on which configuration X is significantly better than Y according to the one-tailed WMW test. For example, the configuration B_{small} performs better than A_{small} on 9 instances according to the WMW test. Tables 2.11, 2.12 and 2.13 show the results obtained by the KIH algorithm, the SAT solver, and the various approaches from the literature. The columns of the tables are arranged such that the results are presented in chronological order, from the two algorithms GGA and LS by Lewis and Paechter [LP07] to the most recent algorithm TSA by Ceschia et al. [CDGS12]. For

Table 2.9 Results of Experiment 2: relative performance of the KIH configurations from Table 2.7

(a) Relative performance of the configurations A_{small}, ..., D_{small}, E on instances small_1, ..., small_20

	A_{small}	B_{small}	C_{small}	D_{small}	E
A_{small}	–	5	6	7	5
B_{small}	9	–	11	10	5
C_{small}	4	5	–	6	6
D_{small}	1	4	5	–	6
E	0	1	10	11	–

(b) Relative performance of the configurations A_{med}, ..., D_{med}, E on instances med_1, ..., med_20

	A_{med}	B_{med}	C_{med}	D_{med}	E
A_{med}	–	6	4	2	11
B_{med}	2	–	3	3	7
C_{med}	1	0	–	1	8
D_{med}	0	1	1	–	7
E	5	6	5	5	–

(c) Relative performance of the configurations A_{big}, ..., D_{big}, E on instances big_1, ..., big_20

	A_{big}	B_{big}	C_{big}	D_{big}	E
A_{big}	–	4	0	0	9
B_{big}	1	–	0	0	7
C_{big}	4	2	–	0	6
D_{big}	7	5	6	–	10
E	6	4	4	3	–

For each pair (X, Y) of configurations, the corresponding table entry indicates the number of instances on which configuration X is superior to configuration Y

each solver from the literature we give the best and average distance to feasibility and, if available, the average runtime in seconds. The SAT solver either found a feasible solution or hit the timeout, i.e., produced no result. On the small, medium, and large instances, we ran the KIH with configurations B_{small}, A_{med}, and D_{big}, respectively. The mean and standard deviation of the average number of room assignments performed per second were determined to be 36150 ± 18856, 24332 ± 18216 and 6333.3 ± 3155.5, on the small, medium and big instances, respectively.

Results

According to the data shown in Table 2.7, the best KIH configurations for the small instances found by I-RACE use the LSD/SO or LSD/RO preprocessing heuristics. This is consistent with our findings from Experiment 1 shown in Table 2.3 which indicate that these heuristics leave the fewest events unscheduled among the preprocessing heuristics under consideration. On the medium and big instances, all tuned configurations use the LSDLD/SO preprocessing. On these instance sets, the results shown in Table 2.3 do not clearly indicate that LSDLD/SO is a superior choice. WMW tests however reveal this is indeed the case: For example, LSDLD/SO is better than LSD/SO on 13 medium instances and 12 big instances. On the other hand, LSD/SO is better than LSDLD/SO on 6 medium and 7 big instances. A similar comparison of LSDLD/SO and LSD/RO is also in favor of the LSDLD/SO heuristic. Considering this, we can conclude that the best overall performance of the KIH is produced with the best preprocessing heuristic.

The tuned configurations indicate that for the remaining parameters ℓ, m, k, and λ, the best configurations can be found within tighter intervals than the ones explored by I-RACE. The parameters in the 16 best configurations identified by irace are shown in Table 2.10.

From the relative performance of the different configuration shown in Table 2.9 we cannot conclude that the tuned configurations found by I-RACE generally show significantly better performance than configuration E from [MW10b*]. However, configurations B_{small}, A_{med}, and D_{big} are clearly superior choices compared to configuration E on the respective instance sets. We further observed that the

Table 2.10 KIH parameter domains of the best performing configurations found by I-RACE

Parameter	Values
ℓ	$[11, 32]$
m	$[2, 3]$
k	$[1, 5]$
λ	$[1.15, 1.72]$

tuned configurations outperform configuration E on similar sets of instances and vice versa: Each of the configurations A_{med}–D_{med} outperforms configuration E on the six instances `med_6`, `med_8`, `med_11`, `med_14`, `med_17`, and `med_19`, and E outperforms every tuned configuration on `med_1`–`med_4`, and `med_10`. Similarly, configurations A_{big}–D_{big} all outperform E on the instances `big_6`, `big_7`, `big_14`, `big_15`, `big_17`, and `big_19`. In turn, E performs better than all tuned configurations on `big_1` and `big_3`. All instances on which significant performance differences were found are part of the reduced instance set used by I-RACE. Therefore, the fact that the configuration E consistently outperforms the tuned configurations on some instances cannot be attributed to the exclusion of instances from the tuning.

Compared to other approaches published in the literature, the KIH performs best on the small and medium instances with respect to the total average distance to feasibility. The KIH was the first published approach that produced feasible timetables on every of the small and medium instances. Overall, on these instances, the KIH uses less CPU time than the HSA, and, with several exceptions, also less than the clique-based approach. However, small differences in runtime should not be considered significant since they may be due to platform and implementation differences. On the big instances, the performance relative to the other approaches is not as good. This indicates scaling issues, which are partly caused by the excessive use of the bipartite matching algorithm: On the big instances, on average only a quarter of the room assignments can be performed per second compared to the medium instances. Thus, the extra time of 600 s for solving the big instances is completely devoted to solving room assignment subproblem instances. Thus, there is no time left for dealing with other sources of the increasing complexity. Since the results shown in Tables 2.11, 2.12 and 2.13 were obtained using the tuned configurations, we cannot expect further substantial improvements without modifying the algorithm.

The SAT solver `minisat` has been employed in order to check if the use of specialized heuristics is justified for finding feasible timetables on the Lewis and Paechter instances. The SAT solver was able to find feasible timetables on 16 small instances, 12 medium instances and 6 big instances using the CNF encoding from Sect. 2.4.3. Whenever the SAT solver was able to solve an instance, the result was produced consistently within a very competitive amount of CPU time. The results indicate that the small instances have become too easy, and benchmarking should focus on the medium and big instances, as advocated by [CDGS12]. On the latter instance sets, the use of specialized heuristics is well justified according to our experiments. A drawback of the SAT-based approach is that if the SAT solver fails to find a satisfying assignment then we get no solution at all. As a remedy, a (partial) MAX-SAT-based approach can be used, which has been demonstrated to produce some of the known best results on the CB-CTT problem by [AAN12].

Table 2.11 Results obtained by the KIH, the SAT solver, and various approaches from the literature on the small instance set from [LPb]

	GGA	LS	HSA		KIH		CB		Minisat
Instance	Best (avg)	Best (avg)	Best (avg)	CPU (s)	Best (avg)	CPU (s)	Best (avg)	CPU (s)	CPU (s)
small_1	0 (0)	0 (0)	0	0	0 (0.00)	0.00	0	0.05	0.44
small_2	0 (0)	0 (0)	0	0	0 (0.00)	0.00	0	0.02	0.50
small_3	0 (0)	0 (0)	0	9	0 (0.00)	0.16	0	1.25	0.47
small_4	0 (0)	0 (0)	0	0	0 (0.00)	0.08	0	0.05	0.46
small_5	0 (1.05)	0 (0)	0	5	0 (0.00)	0.34	0	4.20	0.57
small_6	0 (0)	0 (0)	0	0	0 (0.00)	0.00	0	0.02	0.46
small_7	0 (0)	0 (0)	0	0	0 (0.00)	0.11	0 (0.2)	0.02	0.50
small_8	0 (6.45)	0 (1)	0 (1.9)	79	0 (0.75)	25.01	0 (0.3)	0.05	Timeout
small_9	4 (2.5)	0 (0.15)	0 (3.85)	84	0 (0.00)	3.92	0 (0.15)	0.02	Timeout
small_10	0 (0.1)	0 (0)	0	15	0 (0.00)	0.61	0	1.25	0.79
small_11	0 (0)	0 (0)	0	0	0 (0.00)	0.00	0	0.02	0.50
small_12	0 (0)	0 (0)	0	0	0 (0.00)	0.00	0	2.75	0.71
small_13	0 (1.25)	0 (0.35)	0 (1)	15	0 (0.05)	4.68	0	1.00	0.78
small_14	3 (10.5)	0 (2.75)	3 (5.95)	136	0 (0.65)	24.10	0 (0.7)	70.55	Timeout
small_15	0 (0)	0 (0)	0	0	0 (0.00)	0.12	0	0.55	0.43
small_16	0 (0)	0 (0)	0	13	0 (0.00)	0.13	0	0.55	0.48
small_17	0 (0.25)	0 (0)	0	13	0 (0.00)	0.29	0	2.00	0.62
small_18	0 (0.7)	0 (0.2)	0 (0.45)	36	0 (0.00)	0.20	0 (0.7)	51.65	Timeout
small_19	0 (0.15)	0 (0)	0 (1.2)	25	0 (0.00)	0.87	0	1.00	0.66
small_20	0 (0)	0 (0)	0	0	0 (0.00)	0.00	0 (0.15)	70.85	104.6

For each approach except the SAT solver we give the minimum and average distance to feasibility and, if available, the average CPU time in seconds. For the SAT solver minisat, we give the time taken to find a feasible timetable

Table 2.12 Results obtained by the KIH, the SAT solver, and various approaches from the literature on the medium instance set from [LPb]

	GGA	LS	HSA		KIH		CB		TSA	Minisat
Instance	Best (avg)	Best (avg)	Best (avg)	CPU (s)	Best (avg)	CPU (s)	Best (avg)	CPU (s)	Best (avg)	CPU (s)
med_1	0 (0)	0 (0)	0 (0.00)	0	0 (0.00)	1.23	0 (0.00)	5.85	0 (0.00)	2.00
med_2	0 (0)	0 (0)	0 (0.00)	0	0 (0.00)	1.07	0 (0.00)	1.5	0 (0.00)	1.93
med_3	0 (0)	0 (0)	0 (0.00)	8	0 (0.00)	1.78	0 (0.00)	5.05	0 (0.00)	3.02
med_4	0 (0)	0 (0)	0 (0.00)	3	0 (0.00)	1.18	0 (0.00)	2.05	0 (0.00)	4.03
med_5	0 (3.95)	0 (0)	0 (0.00)	85	0 (0.00)	13.74	0 (0.00)	59.75	0 (0.00)	Timeout
med_6	0 (6.2)	0 (0)	0 (0.00)	20	0 (0.00)	4.25	0 (0.00)	7.95	0 (0.90)	Timeout
med_7	34 (41.65)	14 (18.05)	1 (4.15)	440	0 (0.00)	119.58	0 (3.55)	134.45	0 (0.00)	Timeout
med_8	9 (15.95)	0 (0)	0 (0.00)	12	0 (0.00)	6.46	0 (0.00)	11.35	0 (0.30)	Timeout
med_9	17 (24.55)	2 (9.7)	0 (4.90)	269	0 (0.35)	143.04	0 (2.15)	123.2	0 (0.35)	Timeout
med_10	0 (0)	0 (0)	0 (0.00)	0	0 (0.00)	1.038	0 (0.00)	0.35	0 (0.00)	1.86
med_11	0 (3.2)	0 (0)	0 (0.00)	25	0 (0.00)	0.84	0 (0.00)	3.4	0 (0.00)	2.55
med_12	0 (0)	0 (0)	0 (0.00)	54	0 (0.25)	50.00	0 (0.00)	6.5	0 (0.60)	2.09
med_13	3 (13.35)	0 (0.5)	0 (0.50)	172	0 (0.00)	2.00	0 (0.00)	9.2	0 (0.05)	Timeout
med_14	0 (0.25)	0 (0	0 (0)	59	0 (0.00)	0.25	0 (0.00)	10.9	0 (0.00)	2.26
med_15	0 (4.85)	0 (0)	0 (0.05)	72	0 (0.00)	1.48	0 (0.00)	7	0 (0.00)	25.05
med_16	30 (43.15)	1 (6.4)	1 (5.15)	733	0 (0.00)	7.38	0 (0.30)	21.65	0 (0.00)	Timeout
med_17	0 (3.55)	0 (0)	0 (0.0)	39	0 (0.00)	0.95	0 (0.00)	1.8	0 (0.15)	3.02
med_18	0 (8.2)	0 (3.1)	0 (6.05)	529	0 (0.00)	5.10	0 (0.00)	8.65	0 (0.30)	1.88
med_19	0 (9.25)	0 (3.15)	0 (5.45)	511	0 (0.00)	10.20	0 (0.00)	16.46	0 (0.50)	Timeout
med_20	0 (2.1)	3 (11.45)	2 (10.6)	457	0 (0.00)	8.06	0 (0.65)	24.55	0 (0.55)	10.1

For each approach except the SAT solver we give the minimum and average distance to feasibility and, if available, the average CPU time in seconds. For the SAT solver minisat, we give the time taken to find a feasible timetable

Table 2.13 Results obtained by the KIH, the SAT solver, and various approaches from the literature on the big instance set from [LPb]

	GGA	LS	HSA		KIH		CB		TSA	Minisat
Instance	Best (avg)	Best (avg)	Best (avg)	CPU (s)	Best (avg)	CPU (s)	Best (avg)	CPU (s)	Best (avg)	CPU (s)
big_1	0 (0)	0 (0)	0 (0.00)	0	0 (0.05)	49.30	0 (0.00)	68.4	0 (0.15)	13.93
big_2	0 (0.7)	0 (0)	0 (0.00)	283	0 (0.00)	43.33	0 (0.00)	94.2	0 (0.60)	20.63
big_3	0 (0)	0 (0)	0 (0.00)	447	0 (0.00)	37.21	0 (0.00)	55.4	0 (1.45)	32.96
big_4	30 (32.2)	8 (20.5)	0 (0.00)	406	0 (0.45)	427.17	0 (0.00)	133.5	0 (0.00)	Timeout
big_5	24 (29.15)	30 (38.15)	0 (1.10)	743	3 (9.70)	800.00	1 (3.20)	353.1	0 (0.00)	Timeout
big_6	71 (88.9)	77 (92.3)	5 (8.45)	893	24 (35.15)	800.00	10 (15.40)	319.25	1 (2.85)	Timeout
big_7	145 (157.3)	150 (168.5)	47 (58.30)	966	103 (108.60)	800.00	39 (46.65)	383.85	21 (29.25)	Timeout
big_8	30 (37.8)	5 (20.75)	0 (0.00)	210	0 (1.30)	701.88	0 (0.00)	136.75	0 (0.00)	Timeout
big_9	18 (25)	3 (17.5)	0 (0.05)	419	0 (0.75)	585.97	0 (0.00)	122.6	0 (0.00)	Timeout
big_10	32 (38)	24 (39.95)	0 (1.25)	660	4 (7.60)	800.00	0 (1.95)	319.6	0 (0.00)	Timeout
big_11	37 (42.35)	22 (26.05)	0 (0.35)	444	9 (12.65)	800.00	0 (2.35)	271.7	0 (0.00)	Timeout
big_12	0 (0.85)	0 (0)	0 (0.00)	240	0 (0.00)	23.75	0 (0.00)	54.55	0 (1.15)	20.26
big_13	10 (19.9)	0 (2.55)	0 (0.00)	274	0 (0.00)	83.17	0 (0.00)	84.15	0 (1.15)	146.70
big_14	0 (7.25)	0 (0)	0 (0.00)	271	0 (0.00)	29.40	0 (0.00)	67.8	0 (1.20)	21.68
big_15	98 (113.95)	0 (10)	0 (0.00)	255	0 (3.35)	756.12	0 (0.00)	83.95	1 (3.50)	Timeout
big_16	100 (116.3)	19 (42)	0 (2.00)	755	0 (0.00)	72.74	0 (0.00)	46.85	0 (0.00)	Timeout
big_17	243 (266.55)	163 (174.9)	76 (89.90)	998	0 (10.80)	798.61	0 (2.05)	554.35	12 (22.00)	Timeout
big_18	173 (194.75)	164 (179.25)	53 (62.60)	764	0 (0.00)	280.12	0 (1.70)	437.95	8 (13.55)	Timeout
big_19	253 (266.65)	232 (247.35)	109 (127.00)	998	124 (137.80)	800.00	40 (53.20)	410.45	37 (52.85)	Timeout
big_20	165 (183.15)	149 (164.15)	40 (46.70)	827	6 (9.95)	800.00	9 (14.05)	370.5	11 (15.05)	Timeout

For each approach except the SAT solver we give the minimum and average distance to feasibility and, if available, the average CPU time in seconds. For the SAT solver minisat, we give the time taken to find a feasible timetable

Discussion

Experiment 2 demonstrated the very good performance of the KIH on the small and medium instance sets. As a weakness of the approach, we identified scaling issues of the KIH algorithm on the big instances. These scaling issues result, to some extent, from the algorithm solving room assignment problems almost exclusively. In order to resolve these issues, the use of the bipartite matching algorithm needs to be restricted. This can possibly be achieved by the following two modifications, which can be considered in the future. First, other neighborhood structures could be added, which are not based on the Kempe-exchange. This would help to limit the excessive usage of the bipartite matching algorithm. Second, following a conflict-minimization approach with extra timeslots similar to the one from [CDGS12] seems to be a promising direction. The KIH could be used in order to move events from the extra timeslots to the regular timeslots.

Chapter 3
Fairness in Academic Course Timetabling

In the previous chapter, we have considered search and reconfiguration problems in the context of academic timetabling. That is, all feasible timetables were considered to be of the same quality. For practical purposes, there are typically many more issues that need to be considered. For instance, certain ways of assigning rooms may be preferable to others and spreading each student's courses in a nice way across the week will be appreciated by students and lecturers. Thus, we are typically dealing with an optimization problem, i.e., the task is to find among all feasible timetables one that is best with respect to these additional preferences. These preferences are formalized in terms of functions referred to as *soft constraints*. Each function assigns a value to a given timetable, indicating how desirable or undesirable the timetable is with respect to the corresponding preference. Timetabling problems are typically stated as minimization problems and thus, the values of the soft constraints are interpreted as *penalties*. The penalties are aggregated by the *objective function*, for example, by taking their (weighted) sum. So, more formally, the typical task is to find a feasible timetable that minimizes the aggregated penalty. We will argue shortly, that minimizing the aggregated penalty does not reflect how the stakeholders, typically students and lecturers, judge the quality of a timetable. We propose as a remedy two problem formulations taking the penalties related to the individual stakeholders and, in particular, the penalty distribution over the stakeholders into account.

We can observe that for each stakeholder only a certain part of the overall timetable is relevant. For example, the relevant courses for a student are determined by the curriculum, and for a lecturer the course timetable determines when he or she has to teach. So the perceived quality of a timetable depends largely on the individual timetables of each stakeholder. This is quite a different perspective on the quality of a course timetable compared to using the aggregated penalty as a quality measure. A closely related issue has been noted in the literature, for example, by Bullnheimer, who has argued that minimizing the sum of the penalties "might yield hard consequences for some students" [Bul98]. These "hard consequences"

M. Mühlenthaler, *Fairness in Academic Course Timetabling*, Lecture Notes in Economics and Mathematical Systems 678, DOI 10.1007/978-3-319-12799-6_3

are due to an unequal distribution of the penalty over the stakeholders. That is, minimizing the overall penalty does not prevent situations in which a large part of the penalty hits only a small group of the stakeholders, who would therefore receive a poor timetable in comparison to others. To address these concerns we propose two variants of the course timetabling problem that include fairness criteria in the objective function. Both variants are based on the Curriculum-based Course Timetabling (CB-CTT) problem [DGMS07], which is a popular problem formulation for benchmarking. The two proposed problems, referred to as Max–min Fair Curriculum-based Course Timetabling (MMF-CB-CTT) problem and Jain's Fairness Index Curriculum-based Course Timetabling (JFI-CB-CTT) problem, differ with respect to the underlying notion of fairness.

We further propose an optimization algorithm based on Simulated Annealing (SA) for the MMF-CB-CTT problem, introduce and analyze the Generalized Lexicographic Bottleneck Optimization Problem (GLBOP), which is related to the MMF-CB-CTT problem, propose a decomposition of the MMF-CB-CTT problem and present experimental results for both fair course timetabling problems on standard benchmarking instances.

3.1 Background

In this section, we will provide relevant information on fairness in the context of resource allocation and three combinatorial problems, the CB-CTT problem, the assignment problem, and the Generalized Balanced Academic Curriculum Problem (GBACP). All of these problems are resource allocation problems. Fairness issues have been studied in different disciplines, ranging from social choice theory to engineering and computer science, and a large body of research is available. For a succinct but rather complete overview of the different fairness concepts and how they are related the reader is referred to [LKCS10, Appendix B]. We take these concepts and apply them in the context of course timetabling. Our starting point will be the CB-CTT problem, which is a popular problem formulation for benchmarking [DGMS07, DGS]. The assignment problem appears as a subproblem of the CB-CTT and will be important for the decomposition result in Sect. 3.5. In the last part of this section, we will discuss how achieving a *balanced* allocation of courses to curricula is related to fairness.

3.1.1 Fairness and Resource Allocation

Fairness issues typically arise when scarce resources are allocated to a number of stakeholders with demands. Depending on the context, different notions of fairness may be appropriate. The topic of fairness and how to measure it has received great attention, for example, in the area of economics, where the distribution of

Table 3.1 Fairness in the scheduling and bandwidth allocation literature

Publication	Application	Stakeholders	Resources	Fairness concept
[JCH84]	Bandwidth allocation	Network users	Bandwidth	Jain's fairness index
[DVY03]	Peer-to-peer networks	Peers	Bandwidth	Proportional fairness
[SK08]	Aircraft scheduling	Airlines	Cost	Dispersion measures, bottleneck
[CdMRLS11]	Personnel scheduling	Nurses	Penalty	(Iterated) bottleneck
[MW12*]	Course timetabling	Curricula	Penalty	Jain's fairness index, max-min fairness
[SMO+12]	Personnel scheduling	Nurses	Penalty	Bottleneck

wealth and income is of interest [FS06]. In computer science, fairness aspects have been studied, for example, in the design of network communication protocols, in particular in the context of bandwidth allocation and traffic shaping [LKCS10, JCH84]. Table 3.1 contains a (non-exhaustive) list of papers from the scheduling and bandwidth allocation literature that deal explicitly with fairness. For each entry, we give the corresponding application, the stakeholders, the allocated resources, and a fairness concept. We will cover the different fairness concepts in the subsequent sections, except for proportional fairness, which has received attention mostly in the networking context, see e.g. [KMT98, UK11]. In the context of resource allocation in general, fairness has been studied in [BFT11, Ogr10]. Approximation algorithms for fair optimization problems have been studied in [KRT01, KK06].

Consider a resource allocation problem with n stakeholders. Each feasible resource allocation (feasible solution) s (e.g., a timetable τ in the sense of Definition 2.1) induces a tuple $\mathbf{x} = (x_1, \dots, x_n)$ with non-negative rational entries, where each entry x_i, $1 \leq i \leq n$, corresponds to the how desirable or undesirable the resource allocation s is for stakeholder i. Depending on the context, an entry of $\mathbf{x}$ is either referred to as utility, which is desirable, or cost/penalty, which is undesirable. Thus, we refer to $\mathbf{x}$ either as *utility tuple* or as *penalty tuple*. In [LC10], two main approaches to introducing fairness to a resource allocation problem are identified, referred to as *continuous* and *binary*:

- The *continuous* approach uses a fairness measure to quantify how fair an allocation is. A fairness measure is a function $\zeta : X \to \mathbb{R}$, where X is the set of utility/penalty tuples induced by the feasible allocations.
- The *binary* approach assigns to each solution a binary fairness status. Typically, fair solutions are given in terms of optimal solutions to specific optimization problems. The task is to find a feasible allocation that maximizes (minimizes) a utility (penalty) function. The utility (penalty) function is chosen such that the optimal solutions exhibit a fairness property such as max-min fairness.

Fairness and Efficiency

An allocation is efficient, if it makes best possible use of the available resources. That is, similar to the binary approach to fairness above, an efficient allocation maximizes a certain utility function. In a bandwidth allocation context for example, a possible efficiency metric is the total network throughput. In this scenario, an allocation is efficient if it maximizes the throughput. A more general characterization of efficient allocations is the *Pareto-optimality* property. Informally, an allocation is Pareto-optimal, if any increase (decrease) of an entry of the corresponding utility (penalty) tuple leads to a decrease (increase) of another entry. In the following, we will give a formal definition of weak Pareto-dominance and Pareto-optimality in the *penalty*-scenario because this scenario is relevant for creating fair course timetables. The corresponding definitions for the utility scenario can be derived in a straight-forward manner by replacing "$\leq$", "$<$", and "minimal" by "$\geq$", "$>$", and "maximal", respectively.

Definition 3.1 (Pareto-Dominance) Let $\mathbf{x}, \mathbf{y} \in \mathbb{R}^n$ for some $n \in \mathbb{N}$. The tuple $\mathbf{x}$ *weakly dominates* $\mathbf{y}$, denoted by $\mathbf{x} \preceq_P \mathbf{y}$, if and only if for all $i \in \{1, \ldots, n\}$, $x_i \leq y_i$. Similarly, $\mathbf{x}$ *strictly dominates* $\mathbf{y}$, $\mathbf{x} \prec_P \mathbf{y}$, if and only if $\mathbf{x} \preceq_P \mathbf{y}$ and there is some $i \in \{1, \ldots, n\}$ such that $x_i < y_i$.

The relation $\preceq_P$ is a partial order on $\mathbb{R}^n$, i.e., it is reflexive, transitive and antisymmetric. An inclusion of a set $X \to \mathbb{R}^n$ induces a partial order on X, which we will denote by $(X, \preceq_P)$. For example, using the "obvious" inclusion into $\mathbb{R}^n$, the n-tuples of floating point numbers available in a computer are ordered by $\preceq_P$.

Definition 3.2 (Pareto-Optimality) An element $\mathbf{x} \in X \subseteq \mathbb{R}^n$ is *Pareto-optimal* (or *Pareto-efficient*) if it is minimal in $(X, \preceq_P)$.

Figure 3.1 illustrates the dominance relation on the points $X = \{a, b, c, d, e\}$, which are mapped to $\mathbb{R}^2$. Each point dominates the area above and right of the point. In particular, the point b dominates any point contained in the shaded area, that is, c and e. The points a, b, and d are not dominated by any point in X and therefore they are Pareto-optimal.

Note that for any $\mathbf{w} \in \mathbb{R}^n_{\geq 0}$, if $\mathbf{x} \in X \subseteq \mathbb{R}^n$ minimizes $\sum_i w_i x_i$ then $\mathbf{x}$ is Pareto-optimal. Different fairness notions consider efficiency to different extents. If

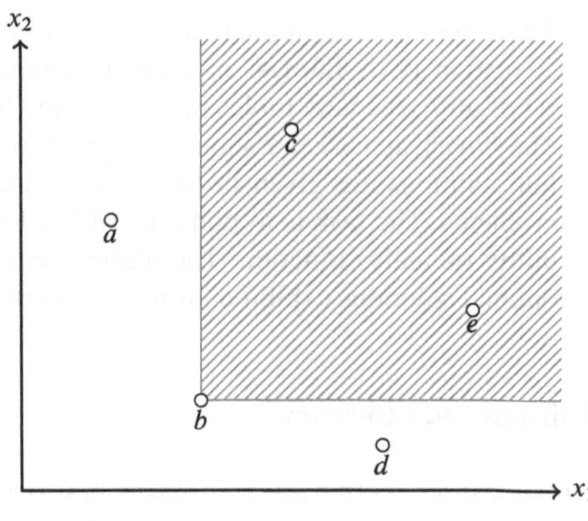

Fig. 3.1 Illustration of the Pareto-dominance relation. The points a, b, and d are not dominated by any other point and are therefore Pareto-optimal

a fair allocation is not Pareto-optimal then a tradeoff between fairness and efficiency arises. The tradeoff can be specified explicitly in the utility/penalty function as shown, for example, in [UK11].

In the resource allocation context, we say that an allocation is Pareto-optimal if its corresponding penalty tuple is Pareto-optimal with respect to the set of penalty tuples induced by all feasible resource allocations.

Max-Min Fairness

Max-min fairness can be stated as iterated application of Rawls' Second Principle of Justice [Raw99]:

> "Social and economic inequalities are to be arranged so that they are to be of greatest benefit to the least-advantaged members of society." (the Difference Principle)

Once the outcome for the least-advantaged group is as good as possible, the difference principle can be applied again to everyone except the least-advantaged group in order to achieve the best-possible outcome for the second least-advantaged group, and so on. The resulting allocation is called max-min fair.

We give a formalization of max-min fairness that is analogous to the one in [KRT01]. We adapt the formalism slightly to suit our penalty allocation scenario. For any tuple $\mathbf{x}$, we denote by $\overleftarrow{\mathbf{x}}$ the tuple containing the entries of $\mathbf{x}$ arranged in non-increasing order. In the following, let $X \subseteq \mathbb{R}^n$ for some $n \in \mathbb{N}$. A tuple $\mathbf{x} \in X$ is at least as good as a tuple $\mathbf{y} \in X$ with respect to max-min fairness, denoted by $x \preceq y$, if and only if $\overleftarrow{\mathbf{x}} \preceq_{lex} \overleftarrow{\mathbf{y}}$, where $\preceq_{lex}$ is the usual lexicographic comparison. The relation $\preceq$ is transitive, and total, over X, but not necessarily antisymmetric. As an example, consider $\mathbf{x} = (1, 2)$ and $\mathbf{y} = (2, 1)$. Then $\mathbf{x} \preceq \mathbf{y}$ and $\mathbf{y} \preceq \mathbf{x}$ does not imply $\mathbf{x} = \mathbf{y}$. Since the lexicographic comparison is performed on the sorted penalty tuples, max-min fairness does not discriminate between stakeholders. The comparison $\preceq$ is purely qualitative, that is, it does not quantify the fairness difference between two allocations.

Definition 3.3 (Max-Min Fairness) A tuple $\mathbf{x} \in X$ is max-min fair, if it is minimal in $(X, \preceq)$.

If a max-min fair tuple $\mathbf{x} \in X$ exists any permutation of its entries is also max-min fair, if it is contained in X. Due to the iterated application of the difference principle, a Max-min fairness enforces an efficient resource usage to some extent. In particular:

Theorem 3.4 *Any max-min fair point* $\mathbf{x} \in X$ *is Pareto-optimal.*

Proof Let $\mathbf{x}^* \in X$ be max-min fair. Assume for a contradiction that there is some $\mathbf{x} \in X$ such that $\mathbf{x}$ strictly dominates $\mathbf{x}^*$, i.e., $\mathbf{x} \prec_P \mathbf{x}^*$. Then all stakeholders are at least as well-off with penalties $\mathbf{x}$ as they are with penalties $\mathbf{x}^*$ and there is at least one stakeholder whose outcome is better in $\mathbf{x}$. Therefore $\mathbf{x}^*$ cannot be max-min fair. □

A weaker version of max-min fairness results if Rawls' Second Principle of Justice is not applied iteratively, but just once. This means that we are concerned with choosing the best possible outcome for the worst-off stakeholder. In the literature, related optimization problems are referred to as bottleneck optimization problems [EF70, PZ11]. Note that in contrast to max-min fairness, an optimal solution to a bottleneck optimization problem is not necessarily Pareto-optimal. In the context of practical academic timetabling, the use of bottleneck optimization is hard to justify: each stakeholder is guaranteed to be at least as well off as the worst-off stakeholder, but no further improvement is considered.

Jain's Fairness Index

Jain's fairness index is an inequality measure proposed in [JCH84], similar to the well-known Gini index [Gin21]. Generally, a highly unequal distribution of resources is considered unfair. Jain's fairness index is the crucial fairness measure that is used in the famous AIMD algorithm [Chi89] used in TCP Congestion Avoidance. The fairness index is a family of functions $J : \mathbb{R}^n_{\geq 0} \rightarrow \mathbb{R}$, $n \geq 2$. The fairness of a utility tuple $\mathbf{x} \in \mathbb{R}^n_{\geq 0}$ is measured as follows:

$$J(\mathbf{x}) = \frac{\left(\sum\limits_{1 \leq i \leq n} x_i\right)^2}{n \cdot \sum\limits_{1 \leq i \leq n} x_i^2} . \tag{3.1}$$

Jain's fairness index has several useful properties: It can be applied for any $n \geq 2$, it is scale independent (for all $\alpha \in \mathbb{R}_{\geq 0}$, $J(\mathbf{x}) = J(\alpha\mathbf{x})$), it is bounded between 0 and 1, and it has an intuitive interpretation: A higher index corresponds to a fairer allocation. Furthermore, if a total amount r of resources is distributed over n stakeholders and $k < n$ stakeholders receive r/k resources while the remaining stakeholders receive nothing, then the fairness index is k/n. In particular, $J(\mathbf{x}) = 1$ means that all stakeholders receive the same amount of resources in allocation $\mathbf{x}$, and $J(\mathbf{x}) = 1/n$ means that all resources are occupied by a single stakeholder, which is considered to be the most unfair allocation. The scale independence property implies that the fairness index shows no inherent preference for efficient utility tuples.

Fairness in the Academic Setting

In the academic setting, fairness has been considered, to some extent, in the Balanced Academic Curriculum Problem (BACP) proposed in [CM01]. Recall from Sect. 2.2 that the BACP problem consists of allocating courses to teaching terms such that workload and prerequisite constraints are satisfied. The objective is to minimize the sum of the deviations from the average workload for each academic term.

The deviation from the average workload is essentially a statistical dispersion measure [KS60, Chapter 3]. Since the objective is to minimize the sum of the deviations, an optimal course allocation is Pareto-optimal with respect to the individual deviations from the average workloads. Thus, the common argument that penalty-minimizing solutions may be unfair due to an unequal (yet efficient) distribution applies: An optimal solution may result in a small deviation from the average workload for some curricula at the expense of other curricula. In our fair course timetabling problem formulations presented in Sect. 3.2, we try to avoid such situations by using Jain's fairness index and max-min fairness in the objective functions of the corresponding optimization problems. For the GBACP however, some specifics limit these undesirable effects: there are fixed upper and lower limits for the workload per term and we can assume that the total workload for each curriculum is sensible.

3.1.2 *The Assignment Problem*

The assignment problem is a classical problem in combinatorial optimization which appears in many applications, for example personnel scheduling, job scheduling and object tracking, just to name a few. For a comprehensive overview of the body of research on the assignment problem and the applications see [BDM09, Pen07]. Polynomial-time algorithms are known for many variants of the assignment problem. Let $A = B = \{1, \ldots, n\}$, for some $n \in \mathbb{N}$. An assignment of the elements of A to the elements of B is a bijection $\sigma : A \rightarrow B$. Typically, assignment problems are optimization problems, that is, among all bijections from A to B, we are looking for one that is optimal with respect to a certain objective function. In the context of (fair) university course timetabling, we are in particular interested in two variants of the assignment problem, namely the Linear Sum Assignment Problem (LSAP) and the Lexicographic Bottleneck Assignment Problem (LBAP).

Definition 3.5 (LSAP) Given a cost function $c : A \times B \rightarrow \mathbb{N}$, find a bijection $\sigma^* : A \rightarrow B$ that minimizes $\sum_{i=1}^{n} c(i, \sigma(i))$.

There exist various algorithms for solving LSAPs efficiently, including the well-known Hungarian algorithm [PS98, p. 248ff] and network flow algorithms, e.g. [GT89]. In the following, let $T_{\text{LSAP}}(n)$ be the time complexity of solving an LSAP instance with $|A| = |B| = n$.

An assignment $\sigma : A \to B$ is called max-min fair, if for any assignment $\sigma' : A \to B$ we have $\mathbf{c}(\sigma) \preceq \mathbf{c}(\sigma')$, where $\mathbf{c}(\sigma) = (c(i, \sigma(i)))_{i=1,\dots,n}$ is a vector-valued function and $\preceq$ is the max-min fair comparison from section "Max-Min Fairness".

Definition 3.6 (LBAP) Given a cost function $c : A \times B \to \mathbb{N}$, find a max-min fair bijection $\sigma : A \to B$.

A LBAP instance can be transformed into an LSAP instance by scaling the cost values appropriately. This results in an exponential blow-up of the cost values, which may be undesirable in practical applications [BR91]. The LBAP can be solved by solving a sequence of bottleneck and zero-one sum optimization problems [SA98], or via a reduction to the lexicographic vector assignment problem, which belongs to the class of *algebraic assignment problems* [BZ80]. We will generalize the latter approach in the next section in order to solve the GLBOP. By the reduction approach, a given LBAP instance with cost function c can be solved in time $O(k \cdot T_{\mathrm{LSAP}}(n))$, where k is the number of distinct values attained by c [DCPT99]. The reduction approach is the following: For each $j \in \{1, \dots, k\}$, let $e_j \in \mathbb{N}^k$ be the vector whose jth component is 1 and all other components are 0. The cost function c is replaced by a tuple-valued function $\mathbf{c}' : A \times B \to \mathbb{N}^k$ such that $\mathbf{c}'(a, \sigma(a)) = e_j$ if $c(a, \sigma(a))$ is the jth largest value attained by c. An assignment σ that yields a lexicographically minimal cost tuple $\sum_{i=1}^{n} \mathbf{c}'(i, \sigma(i))$ is an optimal solution to the corresponding LBAP.

3.1.3 The CB-CTT Problem

The CB-CTT problem is a particular variant of the University Course Timetabling Problem (UCTP) with timeslot availability requirements, which has been studied in Chap. 2. The CB-CTT problem has been introduced in track three of the International Timetabling Competition 2007 (ITC2007) [DGMS07] and has become a popular formulation for benchmarking. The instances and current results can be found on the website [DGS]. Several variations of the problem formulation used in the competition focusing on different aspects of course timetabling have been proposed in [BDCDGS12]. A characteristic of the curriculum-based model is that event conflicts are largely determined by a curriculum structure, see Sect. 2.1, as opposed to a "post-enrollment" scenario, where students pick their courses explicitly and, as a result of these individual choices, any two courses that have at least one student in common are conflicting.

We will give a complete definition of the CB-CTT problem in order to make this chapter self-contained. An equivalent Integer Linear Programming (ILP)-based formulation of the problem can be found in [LL10]. A CB-CTT instance consists of the basic UCTP entities, events, rooms and timeslots, enriched with the following compound structures and additional data:

- Days and timeslots: Each day is divided into a fixed number of time periods. Each combination of a day and a time period is identified with a timeslot and vice versa. The time periods and days are ordered.
- Courses and curricula: Each course consists of a set of events (also referred to as lectures), is taught by a teacher and has a number of students attending the course. The set of courses is a partition of the events. A curriculum is a set of courses.
- Teachers: For each teacher, a set of timeslots is given, in which he or she is available for teaching.
- Rooms: Each room has a capacity, which is the total number of seats.

The course, curriculum, and teacher data contains an implicit description of the conflict graph: Any two lectures of a course are conflicting, any two lectures taught by the same teacher are conflicting, and any two lectures belonging to two courses in a common curriculum are conflicting. The timeslot availability function is given by the teacher availability. Thus, a UCTP instance can be constructed from a CB-CTT instance in a straight-forward fashion. A solution to a CB-CTT instance is an assignment of the lectures to resources (rooms and timeslots). A timetable is feasible if it is conflict-free, overlap-free and satisfies all timeslot availability requirements.

The CB-CTT problem formulation from the ITC2007 features the following four soft constraints:

S1 *RoomCapacity*: Each lecture should be assigned to a room of sufficient size. If the room capacity is not sufficient to accommodate all students then each student without a seat counts as one violation.

S2 *MinWorkingDays*: The individual lectures of each course should be distributed over a certain minimum number of days. Each day below the minimum number of days is counted as one violation.

S3 *IsolatedLectures*: For each curriculum, for each event related to the curriculum there should be another event from the same curriculum in an adjacent timeslot. The timeslot adjacency relation is given by the ordering of the time periods. Each isolated lecture counts as one violation.

S4 *RoomStability*: All lectures of a course should be held in the same room. Each additional room used by a course counts as one violation.

Let $f_{S1}, \ldots, f_{S4}$ be functions that map timetables to violations of the soft constraints S1–S4. Each soft constraint violation results in a "penalty" for the timetable. The total penalty of a timetable τ is aggregated by the objective function f which just takes the weighted sum of the individual soft constraint violations:

$$f(\tau) = \sum_{1 \le i \le 4} w_i f_{Si}(\tau), \qquad (3.2)$$

where $\mathbf{w} = (w_1, \ldots, w_4) = (1, 2, 5, 1)$. The weights $\mathbf{w}$ determine the relative importance of the soft constraints.

The CB-CTT problem is defined as follows:

Definition 3.7 (CB-CTT Problem)

INSTANCE: A CB-CTT instance as defined above.
TASK: Find a feasible timetable τ^* that minimizes f.

The room assignment problem for events in a single timeslot is a LSAP [LL12]. The penalty of assigning, an event e in the given timeslot to a room r is determined by f, assuming that the resource assignments of the events in the other timeslots are fixed.

3.2 Fair Course Timetabling Problem Formulations

In this section, we propose the fair variants of the CB-CTT problem. We replace the original objective function with an objective function that considers fairness. The first problem formulation, the MMF-CB-CTT problem, uses max-min fairness and falls in the category of "binary" approaches to fairness. Max-min fairness is not a suitable concept for exploring the trade-off between fairness and efficiency on the problem instances, since a max-min fair solution is also efficient. Thus, we additionally propose the JFI-CB-CTT problem, a bi-objective problem, which allows for an exploration of the fairness-efficiency trade-off. The JFI-CB-CTT problem falls into category of "continuous" approaches fairness. We picked the CB-CTT problem as a starting point for creating fair course timetables, because it is a popular formulation for benchmarking and we used an extension of this model for our course timetabling application discussed in Chap. 4.

In order to use the fairness concepts from Sect. 3.1.1, we need determine a penalty tuple from a timetable. The central entities in the CB-CTT problem formulation are the curricula. Therefore, our interest lies in achieving a fair distribution of penalty over the curricula. Depending on the application, different stakeholders can be picked, but conceptually this does not change much. In the following, let $\mathcal{I}$ be a CB-CTT instance with curricula $c_1, c_2, \dots, c_k$ and for $j \in \{1, \dots, k\}$ let $f_{S1}|_{c_j}, \dots, f_{S4}|_{c_j}$ be a restriction of the CB-CTT objective functions $f_{S1}, \dots, f_{S4}$ from Sect. 3.1.3 to the courses in curriculum c_j. That is, $f_{Si}|_{c_j}$ computes the number of violations of constraint Si related to the courses in c_j. The functions

$$f|_{c_j}(\tau) = \sum_{1 \leq i \leq 4} w_i \; f_{Si}|_{c_j}(\tau)$$

determine for a timetable τ the aggregated penalty for curriculum c_j, applying the same weighting as the original objective function f. The following penalty function determines for a given timetable τ the corresponding penalty tuple:

$$A(\tau) = \left(f|_{c_1}(\tau),\; f|_{c_2}(\tau), \dots,\; f|_{c_k}(\tau)\right) \; . \tag{3.3}$$

Note that a timetable that minimizes the CB-CTT objective function f is not necessarily Pareto-optimal with respect to A.

Definition 3.8 (MMF-CB-CTT Problem)

INSTANCE: A CB-CTT instance.
TASK: Find a feasible timetable τ^* such that $A(\tau^*)$ is max-min fair.

In order to explore the trade-off between efficient and fair resource allocation in the CB-CTT model, we additionally propose the JFI-CB-CTT problem, which is based on Jain's fairness index. In [JCH84], the fairness index was proposed for a utility allocation scenario. In order to get meaningful results from the fairness index in our penalty scenario, we need to alter the penalty function (3.3). Consider a penalty tuple $\mathbf{x}$ that distributes all penalty equally among $r < k$ curricula, while the remaining $k - r$ curricula receive no penalty. Then, according to the intuitive interpretation of $J(\mathbf{x}) = r/k$ in [JCH84], r curricula are happy with the allocation. In our situation however, this is not the case: $k - r$ curricula are happy because they receive no penalty at all. Therefore, we need to adapt the fairness index to our penalty-based scenario. For $\mathbf{x} \in \mathbb{R}^n_{\geq 0}$ let $x_{max} = \max_{1 \leq i \leq n} x_i$. The fairness index $\tilde{J}$ of a penalty tuple is defined as follows:

$$\tilde{J}(\mathbf{x}) = J((x_{max}, \ldots, x_{max}) - \mathbf{x}) \ . \tag{3.4}$$

In the example above, the fairness index $\tilde{J}$ becomes $(k - r)/k$, which is in agreement with the intuitive interpretation of Jain's fairness index, since $k - r$ curricula receive no penalty. All the other useful properties of J are preserved by this transformation. So far, the fairness index $\tilde{J}(\mathbf{x})$ is undefined for $\mathbf{x} = (a, \ldots, a)$, $a \in \mathbb{R}^n_{\geq 0}$. In our scenario, distributing the penalty uniformly should be the fairest allocation and we let $\tilde{J}(\mathbf{x}) = 1$, which is consistent with the case $r = k$ above.

The following penalty function measures the inefficiency and the unfairness of a timetable τ:

$$F(\tau) = (f(\tau), 1 - \tilde{J}(A(\tau))) \ . \tag{3.5}$$

Definition 3.9 (JFI-CB-CTT Problem)

INSTANCE: A CB-CTT instance $\mathcal{I}$.
TASK: Find the set of feasible timetables which are Pareto-optimal with respect to the penalty tuples induced by F.

As suggested in [MW12*], fairness conditions can be added in an analogous manner to other variants of educational timetabling problems, such as the post-enrollment course timetabling problem and the exam timetabling problem, as well as the nurse rostering problem. For example, the central entities of interest in the post-enrollment course timetabling problem are the individual students. Therefore, the goal were to achieve a fair distribution of penalty over all students. Once an appropriate penalty function has been defined, we immediately get the corresponding optimization problem with a fairness objective. Similar approaches have been explored for the nurse rostering problem [SMO+12, MOS+13] and the examination timetabling problem [MPMÖ13].

Our proposed problem formulations are concerned with balancing the interests between stakeholders, who are in our case the students. In practice however, there are often several groups of stakeholders with possibly conflicting interest, for example students, lecturers and administration. There are several possible ways for extending the problem formulations above to include multiple stakeholders. For example, a multi-objective optimization approach may be considered, where each objective captures the fairness with respect to a particular stakeholder. When using inequality measures like Jain's fairness index for different groups of stakeholders, the inequality values can be aggregated, for instance using a weighted-sum or ordered weighted averaging [Yag88] approach. Furthermore, max-min fairness or a suitable inequality measure can be applied to the different objectives to balance the interests of the different groups of stakeholders.

3.3 Solving the MMF-CB-CTT Problem by Simulated Annealing

SA is a popular local search method proposed by Kirkpatrick et al. [KGV83], which works surprisingly well on many problem domains. SA has been applied successfully to timetabling problems, see for example [Kos05, TD96]. Some of the currently known best solutions to CB-CTT instances from the ITC2007 competition were found by simulated annealing-based methods as shown in [DGS]. The SA algorithm generates a new candidate solution according to some neighborhood exploration method, and replaces the current solution with a certain probability depending on (1) the quality difference (in this context referred to as *energy difference*) between the two solutions and (2) the current temperature. The basic SA procedure is shown in Sect. A.2. The max-min fair comparison $\preceq$ only tells us which of two given solutions is better, but not by how much. Therefore, the main challenge is to find a suitable way to measure the energy difference of two candidate solutions. Using the rank function from Sect. 3.4.2 as a measure of max-min fairness does not lead to satisfactory results in this application. Because of the huge variations in the energy difference magnitudes, it is a very delicate task to determine temperature levels that work well in practice. In the following, we propose three different energy difference measures for max-min fair optimization and provide details on the acceptance criterion, the cooling schedule, and the neighborhood exploration for a complete specification of the algorithm.

We refer to our adaptation of SA to the MMF-CB-CTT problem as MAXMINFAIRSA. The algorithm receives as input a MMF-CB-CTT instance, a feasible timetable, initial and final temperature values and the cooling rate α. The output is a feasible timetable.

Acceptance Probability

Similar to the original SA algorithm, a new solution τ_{next} is accepted if it is at least as good as the current solution τ_{cur}, i.e., $A(\tau_{next}) \preceq A(\tau_{cur})$. If τ_{next} is worse than τ_{cur} then it is accepted with probability

$$\exp\left(-\frac{\Delta E(A(\tau_{cur}), A(\tau_{next}))}{\vartheta}\right) ,$$

where ΔE is the energy difference between the penalty tuples $A(\tau_{cur})$ and $A(\tau_{next})$. We propose three different functions for measuring the energy difference of two timetables: ΔE_{lex}, ΔE_{cw}, and ΔE_{ps}. ΔE_{lex} derives the energy difference from a lexicographic comparison, ΔE_{cw} from the component-wise ratios of the sorted penalty tuples and ΔE_{ps} from the ratios of the partial sums of the sorted penalty tuples. Our experiments presented in Sect. 3.6 indicate that the choice of the energy difference measure has a clear impact on the performance of MAXMINFAIRSA and therefore it is a critical design choice.

For a penalty tuple $\mathbf{x}$, we denote by $\overleftarrow{x}_i$ the ith entry of $\overleftarrow{x}$. The energy difference ΔE_{lex} of two penalty tuples $\mathbf{x}, \mathbf{y} \in \mathbb{R}^n$ is defined as follows:

$$\Delta E_{lex}(\mathbf{x}, \mathbf{y}) = 1 - \frac{1}{n} \cdot \left(\min_{1 \le i \le n} \{i \mid \overleftarrow{y}_i > \overleftarrow{x}_i\} - 1 \right) \tag{3.6}$$

ΔE_{lex} determines the energy difference between $\mathbf{x}$ and $\mathbf{y}$ from the index that determines that $\mathbf{x} \preceq \mathbf{y}$. Therefore, sorted penalty tuples which differ at the most significant indices have a higher energy difference than those differing at later indices. In particular, ΔE_{lex} evaluates to 1 if the minimum is 1, and it evaluates to $1/n$ if the minimum is n.

The energy difference measure ΔE_{lex} considers the earliest index at which two sorted penalty tuples differ but not how much the entries differ. We additionally propose the two energy difference measures ΔE_{cw} and ΔE_{ps} which take this information into account. These energy difference measures are derived from the definitions of approximation ratios for max-min fair allocation problems given in [KRT01]. We need to modify these definition to suit our penalty-based scenario.

Let $\mu_{\mathbf{x},\mathbf{y}}$ be the largest value of the two penalty tuples $\mathbf{x}, \mathbf{y} \in \mathbb{R}^n_{\ge 0}$ offset by a parameter $\delta > 0$, i.e.,

$$\mu_{\mathbf{x},\mathbf{y}} = \max\{\overleftarrow{x}_1, \overleftarrow{y}_1\} + \delta . \tag{3.7}$$

The offset δ has been introduced in order to avoid dividing by zero.

The component-wise energy difference ΔE_{cw} of penalty tuples $\mathbf{x}$ and $\mathbf{y}$ is defined as follows:

$$\Delta E_{cw}(\mathbf{x},\mathbf{y}) = \max_{1 \le i \le n} \left\{ \frac{\mu_{\mathbf{x},\mathbf{y}} - \overleftarrow{y}_i}{\mu_{\mathbf{x},\mathbf{y}} - \overleftarrow{x}_i} \right\} - 1 \tag{3.8}$$

Since all entries are subtracted from $\mu_{\mathbf{x},\mathbf{y}}$, the ratios of the most significant entries with respect to $\preceq$ tend to dominate the value of ΔE_{cw}. Consider, for example, the situation that $\mathbf{y}$ is much less fair than $\mathbf{x}$, say, $\max\{\overleftarrow{\mathbf{x}}_1, \overleftarrow{\mathbf{y}}_1\}$ occurs more often in $\mathbf{x}$ than in $\mathbf{y}$. Then for a small offset δ the energy difference $\Delta E_{cw}(\mathbf{x},\mathbf{y})$ becomes large. On the other hand, if $\mathbf{x}$ is nearly as fair as $\mathbf{y}$ then the ratios are all close to one and thus $\Delta E_{cw}(\mathbf{x},\mathbf{y})$ is close to zero.

The third proposed energy difference measure ΔE_{ps} is based on the ratios of the partial sums $\sigma_i(\mathbf{x})$ of the sorted penalty tuples.

$$\sigma_i(\mathbf{x}) = \sum_{1 \le j \le i} \overleftarrow{x}_j \ .$$

Our intention for using partial sums of the sorted penalties is to give the stakeholders receiving the most penalty more influence on the resulting energy difference compared to ΔE_{cw}. The energy difference ΔE_{ps} is defined as:

$$\Delta\Delta E_{ps}(\mathbf{x},\mathbf{y}) = \max_{1 \le i \le n} \left\{ \frac{i \cdot \mu_{\mathbf{x},\mathbf{y}} - \sigma_i(\overleftarrow{\mathbf{y}})}{i \cdot \mu_{\mathbf{x},\mathbf{y}} - \sigma_i(\overleftarrow{\mathbf{x}})} \right\} - 1 \ . \tag{3.9}$$

Cooling Schedule

The temperature at time t is determined according to a standard geometric cooling schedule:

$$\vartheta(t) = \alpha^t \cdot \vartheta_{max} \ ,$$

where α is the cooling rate. Geometric cooling schedules are widely used in practice and are known to work well in many problem domains including timetabling problems, see for example [LA87, KAJ94, TD98]. In order to make the algorithm behavior more consistent for different termination criteria, instead of specifying the cooling rate, we determine it from ϑ_{max}, the desired minimum temperature ϑ_{min} according to:

$$\alpha = \left(\frac{\vartheta_{min}}{\vartheta_{max}} \right)^{\frac{1}{s}} \ , \tag{3.10}$$

where s is the timeout or maximum iteration count. Thus, at the beginning ($t = 0$), the temperature level is ϑ_{max} and when the timeout or maximum iteration count is reached, $t = s$, then the temperature level becomes ϑ_{min}.

Neighborhood

In our MAXMINFAIRSA algorithm, a neighbor is selected at random from the Kempe-exchange Neighborhood (KX) in each step. After performing a Kempe-exchange, the room assignments in affected timeslots are updated. We will perform the room assignment for each timeslot individually, because the room stability constraint S4 introduces interdependencies between the costs of assigning events to rooms in different timeslots, as noted in [LL12]. Thus, solving a room assignment problem for each timeslot is a heuristic approach to updating the room assignment of the courses in two timeslots. We consider two approaches to performing the room assignment for events in a single timeslot: The first one, models the room assignment as a LSAP, such that the costs are determined by the CB-CTT objective function f. The second one models the room assignment as a GLBOP as shown in Sect. 3.5. We will refer to the corresponding MAXMINFAIRSA variants as MMFSA-LSAP and MMFSA-GLBOP, respectively.

3.4 The Generalized Lexicographic Bottleneck Optimization Problem

In a penalty allocation scenario, combinatorial problems with a max-min fairness objective are often referred to as lexicographic bottleneck optimization problems in the literature. Such problems have been studied for example in [BR91, SA98, GKR12]. Here, we propose and study the Generalized Lexicographic Bottleneck Optimization Problem (GLBOP), which is based on the combinatorial model for max-min fair optimization from [BR91]. The task is to pick elements from a given ground set such that the resulting cost tuple is max-min fair. In the original model, the cost of each element in the ground set is a nonnegative integers. We extend this model by using finite multi-sets of nonnegative integers as cost values. We present a solution approach based on cost scaling that allows a GLBOP to be solved efficiently, if the corresponding min-sum optimization problem can be solved efficiently. In Sect. 3.5, we will model the room assignment subproblem of the MMF-CB-CTT problem as a GLBOP.

3.4.1 Problem Formulation and Properties

Let $E = \{e_1, \ldots, e_m\}$ be a finite ground set and let $\mathcal{S} = \{S_1, S_2, \ldots\}$ be a set of feasible solutions. Each solution $S \in \mathcal{S}$ contains exactly n items from the ground set, that is $S \subseteq E$ and $|S| = n$. As an example, consider E to be the edges of a complete bipartite graph and $\mathcal{S}$ to the set of perfect matchings. Furthermore, let N be a finite set of nonnegative integers. The weight function $w : E \to \mathcal{M}(N)$ assigns to a each item in the ground set a weight, which is a finite multiset of the numbers in N. The weight $W(S)$ of a solution $S \in \mathcal{S}$ is the disjoint union of the individual weights,

$$W(S) = \dot{\bigcup_{e \in S}} w(e) \ .$$

We order the weights $\mathcal{M}(N)$ by a straightforward extension of the max-min fair comparison $\preceq$ from section "Max-Min Fairness". Two weights $w_1, w_2 \in \mathcal{M}(N)$ are compared by arranging the items in each multiset in non-increasing order and performing a lexicographic comparison of the sorted sequences. In contrast to the usual max-min fair comparison, the sorted sequences do not necessarily have the same length. Let Seq(N) be the set of finite sequences over the alphabet N that are arranged in non-increasing order. For two sequences $s, s' \in \text{Seq}(N)$, $s \preceq_{lex} s'$ iff one of the following is true: (1) $s = s'$, (2) s is a prefix of s', (3) there is a decomposition $s = zuv$, $s' = zuw$ such that z is a maximal prefix of s and s', $u, v \in N$ and $u < v$. Due to the conceptual similarity, we will use the symbol $\preceq$ for the comparison with respect to max-min fairness in the new setting as well.

Given a ground set, feasible solutions and a weight function, the task is to find feasible solution S such that $W(S)$ is max-min fair, that is

$$\min_{\preceq} \ W(S) \quad \text{s.t. } S \in \mathcal{S} \ . \qquad \text{(GLBOP)}$$

Note that the minimum weight is determined according to the comparison $\preceq$. If $|w(e)| = 1$ for each $e \in E$ then we have a Lexicographic Bottleneck Optimization Problem (LBOP) as defined in [BR91].

Consider a weight function of the form $W' : \mathcal{S} \to \mathbb{N}$. Then the Min-sum Optimization Problem (SOP) is the following problem:

$$\min_{\leq} \sum_{e \in S} W'(e) \quad \text{s.t. } S \in \mathcal{S} \ . \qquad \text{(SOP)}$$

Let $T_{n,m}$ be the time required for solving the problem (SOP).

Theorem 3.10 *A GLBOP instance can be solved in time $O(|N| \cdot T_{n,m})$.*

Proof Following the vectorial approach of Della Croce et al. in [DCPT99], we reduce the GLBOP to a Lexicographic Vector Optimization Problem (LVOP). Let $t = |N|$ and for $1, \leq i \leq t$, let t_i denote the ith largest item in N. Let the function $f : \mathcal{M}(N) \to \mathbb{N}^t$ assign to a multiset v of the items N a tuple $(v_1, \ldots, v_t) \in \mathbb{N}^t$ such that

$$v_j = \text{mult}_v(t_j), \quad j = 1, \ldots, t \ ,$$

where $\text{mult}_v : N \to \mathbb{N}$, and for each $a \in \mathbb{N}$, $\text{mult}_v(a)$ is the multiplicity of a in v. Now, we have to solve the following problem:

$$\min_{\preceq_{lex}} \sum_{e \in S} f(W(e)) \quad \text{s.t. } S \in \mathcal{S} \ . \tag{LVOP}$$

We show that a solution $S \in \mathcal{S}$ is an optimal solution to (LVOP) if and only if it is an optimal solution to (GLBOP). Suppose for a contradiction that $S \in \mathcal{S}$ is an optimal solution to (LVOP) that is not optimal to (GLBOP). Then there is a solution $S' \in \mathcal{S}$ such that $W(S') \prec W(S)$. As a consequence, $f(W(S')) \prec_{lex} f(W(S))$ by the construction of the cost tuples above. The "only if" part is analogous. This is a contradiction to the optimality of S'. The problem (LVOP) can be solved in time $O(t \cdot T_{n,m})$ because each elementary operation involving the weights is now performed on a t-tuple. □

As noted in [DCPT99], the vectorial approach is essentially cost scaling. The construction of the cost tuples above enables us to naturally handle multisets as cost values. We will see shortly that in the context of the MMF-CB-CTT, if an item of the ground set has a weight of cardinality k, then choosing this item to be part of the solution concerns k different stakeholders.

3.4.2 *Measuring Solution Quality*

There is a wealth of statistical tools available for extracting meaningful information about the absolute and relative performance of randomized heuristic algorithms. In Sect. 2.5 for example, methods of descriptive statistics and statistical inference were used in order to evaluate the Kempe Insertion Heuristic (KIH). Due to the qualitative nature of the max-min fair comparison, so far only statistical tools based on ordinal comparisons can be used for evaluating randomized algorithms for max-min fair optimization problems. We propose a novel approach to partially overcome this limitation. Similar to the (GLBOP) from the previous section, we consider combinatorial problems such that the cost of a feasible solution is a finite multiset. Thus, ordinary lexicographic bottleneck objectives are a special case. The main idea is to construct an isomorphism from the cost multisets ordered by $\preceq$, to an interval of the natural numbers ordered by the usual $\leq$ relation. Using this isomorphism, we can perform computations on natural numbers, and then retrieve the corresponding cost multiset of the result. As a consequence, if we have a set of penalty tuples, then we can determine, for example, an approximation of the "average" penalty tuple. We will use this result for our evaluation of the MAXMINFAIRSA algorithm on standard CB-CTT instances in Sect. 3.6.

Let $k \in \mathbb{N}$ and $N = \{0, \ldots, k\}$. Further, let $\text{Seq}_n(N)$ denote the non-increasing sequences of length n over the alphabet $(N, \leq)$.

Lemma 3.11 *Let* rank *map any non-increasing sequence* $x_1 \geq x_2 \geq \ldots \geq x_n$ *over* N *to a number such that*

$$\text{rank}(x_1, \ldots, x_n) = \sum_{i=1}^{n} \binom{n + x_i - i}{x_i - 1} . \tag{3.11}$$

Then rank *is an isomorphism* $(\text{Seq}_n(N), \prec_{lex}) \rightarrow (\{0, \ldots, \ell\}, <)$, *where* $\ell = |\text{Seq}_n(N)| - 1$.

Proof $(\text{Seq}_n(N), \prec_{lex})$ is a linearly ordered set with least element $(0, \ldots, 0)$. Since $\text{Seq}_n(N)$ is linearly ordered by $\prec_{lex}$, there is a unique number r_s for each $s \in \text{Seq}_n(N)$, which is the cardinality of the set $\{s' \in \text{Seq}_n(N) \mid s' \prec_{lex} s\}$. Thus, the function mapping each $s \in \text{Seq}_n(N)$ to r_s is a bijective mapping and it is order-preserving as required. It remains to be shown that rank(s) computes r_s for all $s \in \text{Seq}_n(N)$.

Let $s = (x_1, \ldots, x_n) \in \text{Seq}_n(N)$. The value r_s can be determined by the following recursion:

$$r_{x_1,\ldots,x_n} = r_{x_2,\ldots,x_n} + r_{x_1,0,\ldots,0} . \tag{3.12}$$

This recursion separately counts the non-increasing sequences $\{s' \in \text{Seq}_n(N) \mid (x_1, 0, \ldots, 0) \preceq_{lex} s' \prec_{lex} s\}$ and $\{s' \in \text{Seq}_n(N) \mid s' \prec_{lex} (x_1, 0, \ldots, 0)\}$. The number of sorted sequences of length n over an ordered alphabet of size k is $\binom{n+k}{k}$. Thus,

$$r_{\underbrace{x_1, 0, \ldots, 0}_{\text{length } n}} = \binom{n + x_1 - 1}{x_1 - 1} .$$

In particular, for $s \in \text{Seq}_1(N)$ we have $r_x = \binom{x}{x-1} = x$. Unfolding the recursion (3.12) yields Eq. (3.11). Therefore, rank(s) computes r_s for each $s \in \text{Seq}_n(N)$. □

The argument above can be extended to non-increasing sequences of length *at most n* by choosing as alphabet $N \cup \{-\infty\}$ and constructing from each sequence of length less than n a sequence of length n by padding it with $-\infty$. Please note that the alphabet can be any finite totally ordered set $(A, <_A)$, since it is isomorphic to $\{0, \ldots, k\}$ for some $k \in \mathbb{N}$.

Let $\mathcal{M}_n(N)$ denote the finite multisets of cardinality n over a finite alphabet $(N, <)$. Similar to the problem (GLBOP), consider an instance of some combinatorial minimization problem with feasible solutions $\mathcal{S}$ and a cost function of the form

$$w : \mathcal{S} \rightarrow \mathcal{M}_n(N) .$$

For $x \in \mathcal{M}_n(N)$ let $\overleftarrow{x}$ be the non-increasing sequence of length n containing the items of x.

Theorem 3.12 *Let $\ell = |\mathcal{M}_n(N)| - 1$. Then the mapping*

$$\rho : (\mathcal{M}_n(N), \prec) \rightarrow (\{0, \ldots, \ell\}, <)$$
$$\rho(x) \mapsto \text{rank}(\overleftarrow{x}) \ ,$$

is an isomorphism.

Proof Each multiset $x \in \mathcal{M}_n(N)$ can be represented in a unique way as a non-increasing sequence of length n. Thus, we can apply the isomorphism rank from Lemma 3.11 to $\overleftarrow{x}$. □

We establish a similar result for utility allocation scenarios. In this context, the fairness of two allocations $x, y \in \mathcal{M}_n(N)$ can be compared as follows: x is fairer than y, denoted by $x \prec_{max} y$, if $\overleftarrow{y} \prec_{lex} \overleftarrow{x}$, where $\overleftarrow{x}$ denotes the non-decreasing sequence of the items in x. Intuitively, the worst-off stakeholder should receive as much as possible, and, under this condition, so should the second worst-off, and so on. Let $m = \max\{N\}, \ldots, \max\{N\} \in \text{Seq}_n(N)$, and let $a - b$ denote the element-wise subtraction of two sequences $a, b \in \text{Seq}_n(N)$.

Theorem 3.13 *Let $\ell = |\mathcal{M}_n(N)| - 1$. Then the mapping*

$$\rho' : (\mathcal{M}_n(N), \prec_{max}) \rightarrow (\{0, \ldots, \ell\}, <)$$
$$\rho'(x) \mapsto \text{rank}(m - \overleftarrow{x}) \ ,$$

is an isomorphism.

Proof Let $x, y \in \mathcal{M}_n(N)$. Then $g : (\mathcal{M}_n(N), \prec_{max}) \rightarrow (\mathcal{M}_n(N), \prec)$, $g(x) = m - \overleftarrow{x}$ is an isomorphism. Thus, $\rho' = (\text{rank} \circ g)(x) = \text{rank}(m - \overleftarrow{x})$ is an isomorphism as required. □

3.5 A Decomposition of the MMF-CB-CTT Problem

It is a common approach to decompose the CB-CTT problem in a way that room and timeslot assignment are preformed separately, see for example [LL12, LH10]. The decomposition isolates, for a single timeslot, the assignment of courses to rooms, from the rest of the problem. Given a feasible timetable, the task is to find an optimal room assignment for a given timeslot, while the rest of the timetable remains fixed. Optimizing the rest of the timetable can be considered the master problem corresponding to the room assignment subproblem for a particular timeslot. The room assignment subproblem of the CB-CTT problem is a LSAP which can be solved efficiently (see Sects. 3.1.2 and 3.1.3). Here, we will establish the following analogous result for the MMF-CB-CTT problem: An optimal room assignment for the courses in a single timeslot can be determined by solving a GLBOP instance, which can also be done efficiently.

In the following, let $\mathcal{I}$ be a MMF-CB-CTT instance, with courses C, rooms R, timeslots P, and curricula $U \subseteq \mathcal{P}(C)$. By C_p we denote the set of courses scheduled in the timeslot $p \in P$. Please note that C_p is determined by the solution of the master problem, i.e. a given timetable. Furthermore, let $U_e = \{u \in U \mid e \in u\}$.

Theorem 3.14 *The room assignment subproblem of the MMF-CB-CTT problem is a GLBOP.*

Proof We construct a GLBOP that models the room assignment subproblem for a fixed timeslot p. We can assume that $|C_p| = |R|$ since if not, we can add suitable dummy nodes to either C_p or R. Let $G = (C_p \cup R, E)$ be a complete bipartite graph, where $E = \{\{e, r\} \mid e \in C, r \in R\}$. The ground set is E and the feasible solutions $\mathcal{S}$ are all perfect matchings in G. Assigning a course $e \in C_p$ to a room $r \in R$ completes the timetable τ from the perspective of all curricula in U_e and therefore determines their cost entries $c(u, \tau)$ for each $u \in U_e$. We denote the cost $c(u, \tau)$ caused by assigning e to r by $c_{e \to r}(u)$. Thus, the weight $w(e, r)$ of an item $\{e, r\} \in E$ is the multiset

$$w(e, r) = \biguplus_{u \in U_e} \{c_{e \to r}(u)\} \ .$$

The cost entries of the curricula that do not contain any course in C_p are not altered and have no influence on the optimality of a particular room assignment. Therefore, the room assignment subproblem of the MMF-CB-CTT problem can be written as:

$$\min_{\preceq} \quad \biguplus_{\{e,r\} \in S} w(e, r) \quad \text{s.t. } S \in \mathcal{S} \ , \tag{3.13}$$

which is a GLBOP. □

Note that according to the MMF-CB-CTT problem formulation, a course can be assigned to any of the rooms. If desired, room availabilities can be added to the model in a straightforward manner: If a room r is unavailable for a particular course e, then the edge $\{e, r\}$ in the ground set is assigned a suitably large weight.

Figure 3.2 shows a simple example instance of a room assignment problem modeled as a GLBOP. There is only a single timeslot p, two courses, $C = C_p = \{c_1, c_2\}$, and two rooms $R = \{r_1, r_2\}$. However, the course e_2 is in two curricula and thus determines the cost entries of two stakeholders in the overall penalty tuple. The cost on each edge connected to e_1 shows the costs generated for each of the two curricula when assigning e_2 to r_1 or r_2. Figure 3.2a, b are LBAP instances that reflect only costs for one of the two curricula of e_2. The assignments highlighted are both optimal solutions to the individual LBAPs, but none of them is optimal for the room assignment shown in Fig. 3.2c.

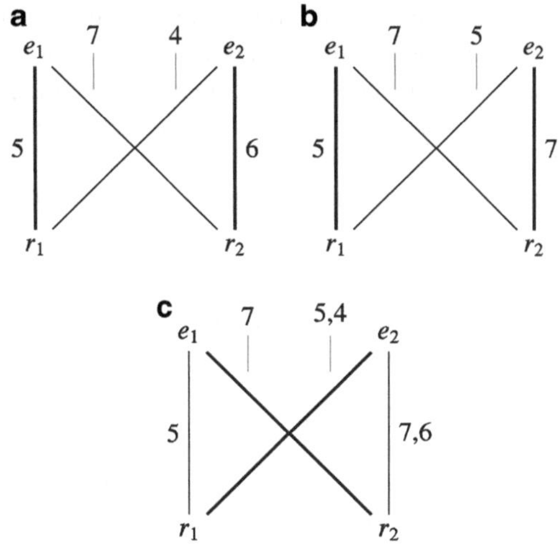

Fig. 3.2 A room assignment problem example with two courses, e_1 and e_2, and two rooms r_1 and r_2. The thick edges are optimal assignments. The shown optimal solutions of the two LBOPs (**b**) and (**a**) are not optimal solutions to the GLBOP instance (**c**). (**a**) LBAP 2: optimal cost 6,5. (**b**) LBAP 1: optimal cost 7,5. (**c**) GLBOP: optimal cost 7, 5, 4

Corollary 3.15 *For a given timeslot p, the room assignment subproblem can be solved in time* $O(|U| \cdot T_{\text{LSAP}}(n))$*, where* $n = \max\{|C_p|, |R|\}$.

Proof Problem (3.13) is an assignment problem, just as the room assignment subproblem of the CB-CTT problem. Only the objective function is different. Hence, combining Theorems 3.10 and 3.14 yields the result. □

Remark 3.16 As noted in [LL12], the room stability constraint (S4) introduces dependencies between room assignments in different timeslots, which prevent us from extending the decomposition to more than a single timeslot.

Remark 3.17 In the general case, whenever there is a decomposition of (SOP) into a master problem and a subproblem, such that the subproblem can be solved efficiently, then the subproblem of the corresponding LBOP can be solved efficiently if it is a GLBOP. This observation may be useful when considering the lexicographic bottleneck variant of a problem of the form (SOP).

In Sect. 3.6 we will provide experimental evidence that solving the room assignment subproblem to optimality is useful for improving the performance of a heuristic algorithm for the MMF-CB-CTT problem.

3.6 Evaluation

We will first address the question how fair the known best timetables for 32 CB-CTT instances are with respect to Jain's fairness index and max-min fairness. We will then evaluate the performance of the MaxMinFairSA algorithm and determine the performance impact of using the decomposition from Sect. 3.5. Finally, we will investigate the tradeoff between fairness and efficiency in the JFI-CB-CTT model.

To aid the presentation of the data, the penalty tuples are given in a sorted multiset form: The penalty values are sorted in non-increasing orders the multiplicities of the values are shown as exponents. Thus, the penalties of the worst-off stakeholders are leftmost. For example, the penalty tuple $(0, 5, 0, 0, 5)$ is represented as $5^2 0^3$. Note that concerning max-min fairness, it is not important to which curriculum a penalty value corresponds, so we omit this information.

3.6.1 Fairness of the Known Best Timetables

We the fairness of the known best timetables to a set of CB-CTT instances with respect to Jain's fairness index and max-min fairness. These timetables were not created with fairness in mind, but the objective was find a timetable which minimizes the function f given in Eq. (3.2). Our evaluation should provide some insight if it is necessary to introduce fairness criteria in the objective function in this setting.

The instances under consideration are the instances `comp01`, ..., `comp21` from the ITC2007 competition, seven instances from the University of Udine, `DDS1`, ..., `DDS7`, and four instances from the FAU school of engineering in Erlangen, `erlangen-2011-2`, `erlangen-2012-1`, `erlangen-2012-2`, and `erlangen-2013-1`. All instances and solutions are available on the website [DGS]. In Table 3.2, τ_{best} refers to the known best solution for the given instance. The column k gives the number of curricula for each instance, which corresponds to the number of stakeholders in the max-min fair problem model. The function f is the CB-CTT objective function. The column $\tilde{J}(\tau_{best})$ contains the fairness index for the individual timetables according to (3.4). The column $\vec{A}(\tau_{best})$ contains the penalty tuples for the individual stakeholders arranged in non-increasing order.

The data indicates that, according to the fairness index $\tilde{J}$, the timetables with a low total penalty are also rather fair. This can be explained by the fact that these timetables do not have a large amount of penalty to distribute over the curricula. Thus, most curricula receive little or no penalty and consequently, the distribution is fair for most curricula. We will show in Sect. 3.6.3, that for solutions to the ITC2007 instances with rather large total penalty there is still some room for improvement with respect to $\tilde{J}$. In the rightmost column of Table 3.2 we observe that sum of the values of a penalty tuple is generally larger than the CB-CTT objective value. The reason for this is that the penalty related to a course is counted for each curriculum the course belongs to. Except for a few instances, the penalty is assigned to only a few curricula while a majority of curricula receives no penalty.

Table 3.2 Fairness of the current best timetables for 32 CB-CTT instances [DGS]

Instance	k	$f(\tau_{best})$	$\tilde{J}(\tau_{best})$	$\overleftarrow{A}(\tau_{best})$
comp01	14	5	0.8571	$5^2, 0^{12}$
comp02	70	24	0.9515	$4, 2^{10}, 0^{59}$
comp03	68	66	0.9114	$13, 10^3, 9, 7^2, 6^4, 5^{13}, 4, 2^6, 0^{37}$
comp04	57	35	0.8964	$7, 6^3, 5^4, 4^2, 2, 0^{46}$
comp05	139	291	0.8277	$41^2, 36^7, 35^5, 32^5, 31^6, 30^9, 28, 27^7, \ldots$
comp06	70	27	0.9657	$12, 7^2, 5^4, 2^3, 0^{60}$
comp07	77	6	0.9870	$6, 0^{76}$
comp08	61	37	0.9020	$7, 6^3, 5^4, 4^2, 2^2, 0^{49}$
comp09	75	96	0.8047	$10^5, 9, 7^{10}, 6^6, 5^{10}, 4, 2, 0^{41}$
comp10	67	4	0.9701	$2^2, 0^{65}$
comp11	13	0	1	0^{13}
comp12	150	300	0.9128	$45, 30^{14}, 28, 27^2, 26^5, 25^{19}, 22^4, 21^6, \ldots$
comp13	66	59	0.8830	$8, 7, 6^5, 5^7, 4^2, 2^3, 0^{47}$
comp14	60	51	0.9023	$8^4, 7, 5^2, 2^6, 0^{47}$
comp15	68	66	0.8495	$10^3, 9^3, 7, 6^4, 5^{13}, 4, 2^7, 0^{36}$
comp16	71	18	0.9176	$7^2, 5^7, 4, 0^{61}$
comp17	70	56	0.9248	$10^2, 6^3, 5^9, 2^4, 0^{52}$
comp18	52	62	0.9009	$17, 15, 14, 13, 11, 10, 9^2, 5^{19}, 2^2, 0^{23}$
comp19	66	57	0.9612	$13, 7, 6^4, 5^2, 4, 2^7, 0^{50}$
comp20	78	4	0.9744	$2^2, 0^{76}$
comp21	78	74	0.8834	$12, 11, 10^4, 9, 7^4, 6^4, 5^{12}, 4, 2^3, 0^{47}$
DDS1	99	48	0.77	$5^{21}, 2^9, 0^{69}$
DDS2	11	0	1	0^{11}
DDS3	9	0	1	0^9
DDS4	105	17	0.97	$15^3, 2, 0^{101}$
DDS5	44	0	1	0^{44}
DDS6	62	0	1	0^{62}
DDS7	37	0	1	0^{37}
erl.2011-2	755	1,171	0.9273	$20, 19^2, 17^5, 16, 15^{14}, 14^8, 12^{39}, 11^5, \ldots$
erl.2012-1	768	943	0.97	$22, 20^5, 17, 15^4, 12^{10}, 10^{19}, 9^8, 8, 7^{40}, \ldots$
erl.2012-2	850	1,354	0.9687	$56, 53^3, 51^{10}, 49^8, 48, 46, 44^7, 19^2, 15^6, \ldots$
erl.2013-1	738	1,092	0.9694	$19, 17, 15^4, 14^2, 12^6, 11^3, 10^6, 9^8, 8^2, \ldots$

For each instance, we give the number k of curricula and the CB-CTT objective value $f(\tau_{best})$ of the known best solution τ_{best}, the fairness index $\tilde{J}(\tau_{best})$, and the sorted penalty values $\overleftarrow{A}(\tau_{best})$

3.6.2 *Evaluation of the* MAXMINFAIRSA *Algorithm*

In the following, we will investigate three aspects of the MAXMINFAIRSA algorithm. First, we will study the impact of the energy difference measures ΔE_{lex}, ΔE_{cw}, and ΔE_{ps} on the solution quality. In our second experiment, we will investigate, if using the decomposition approach from Sect. 3.5 for updating the room assignment in algorithm MAXMINFAIRSA improves the overall performance of the algorithm. Finally, we will compare the best results we obtained against the known best solutions, which were not created with max-min fairness in mind, to check if the improvements are worthwhile.

Recall that the MAXMINFAIRSA algorithm uses the KX neighborhood and updates the room assignment after each Kempe exchange. In our experiments, we distinguish two variants of the algorithm, MMFSA-LSAP and MMFSA-GLBOP, which differ with respect to the room assignment update. The first one is the variant proposed in [MW12*], which models the room assignment subproblem as a LSAP. The second variant was originally proposed in [MW13*]. It relies on the MMF-CB-CTT decomposition from Sect. 3.5, which models the room assignment problem as a GLBOP.

In contrast to the experimental setup in [MW12*], we will not use a timeout, but set a fixed number of iterations for the algorithm MAXMINFAIRSA. The advantages and drawbacks of this choice are discussed in Sect. 2. In practice, solving (GLBOP) takes significantly more time than solving an LSAP. The increase in runtime is due to overhead required by construction of the cost tuples ($O(|U|^2)$) and the $O(|U|)$ factor for solving (LVOP) (see Corollary 3.15), where U is the set of curricula. The additional overhead may also be caused by implementation itself. Therefore, setting a maximum iteration count seemed more appropriate for our investigations since both MAXMINFAIRSA variants are then able to solve the same number of room-assignment subproblems.

Preliminary Screening The energy difference measure ΔE_{cw} has an offset parameter $\delta \in \mathbb{R}_{\geq 0}$, which has been introduced in order to avoid a division by zero. In [MW12*], the performance impact of choosing δ from $\{1, 10^{-2}, 10^{-3}, 10^{-6}\}$ has been investigated. For each choice, 50 independent runs of MAXMINFAIRSA were performed on the ITC2007 competition instances with $\vartheta_{max} = 5$, $\vartheta_{min} = 0.01$, and a timeout of 192 s. The best performing configuration was determined according to the one-sided Wilcoxon-Mann-Whitney (WMW) test with a significance level of 0.01. According to this test, the $\delta = 10^{-3}$ was significantly better than $\delta = 1$ on four instances and better than $\delta = 10^{-6}$ on two instances. Since the overall performance differences were not substantial, we forgo a further investigation and set $\delta = 10^{-3}$ in our computational experiments. Further, we set $\vartheta_{max} = 5$ and $\vartheta_{min} = 0.01$, and determine the cooling rate according to Eq. (3.10).

Experiment 1: Impact of the Energy Difference Measures

We investigate the impact of the three energy difference measures ΔE_{lex}, ΔE_{cw} and ΔE_{ps} on the performance of MMFSA-LSAP and MMFSA-GLBOP.

Experiment Design Our computational experiments were performed on 32 instances, the 21 ITC2007 competition instances, the seven DDS instances and the four Erlangen instances, which are available on the website [DGS]. The energy difference measure is selected from $\{\Delta E_{lex}, \Delta E_{cw}, \Delta E_{ps}\}$. Recall that an energy

Table 3.3 For each choice of two configurations C_1, C_2, the number of instances is shown, on which configuration C_2 outperforms configuration C_2

		C_2					
		MMFSA-LSAP			MMFSA-GLBOP		
C_1		ΔE_{lex}	ΔE_{cw}	ΔE_{ps}	ΔE_{lex}	ΔE_{cw}	ΔE_{ps}
MMFSA-LSAP	ΔE_{lex}	–	0	11	6	5	11
	ΔE_{cw}	28	–	24	6	5	16
	ΔE_{ps}	5	0	–	6	5	8
MMFSA-GLBOP	ΔE_{lex}	23	12	23	–	0	21
	ΔE_{cw}	25	14	24	8	–	21
	ΔE_{ps}	11	2	11	0	0	–

difference measure can be constructed from Theorem 3.12. We did not consider this energy difference measure here, because the energy differences vary over several orders of magnitude and therefore, choosing temperature levels is a delicate task and the results were not satisfactory. For each energy difference function we independently performed 50 runs of MAXMINFAIRSA. In each run, we performed 10^5 iterations of MAXMINFAIRSA. The quality a MAXMINFAIRSA run is given in terms of the penalty tuple of the best timetable found by the algorithm. We used the one-sided WMW test with a significance level of 0.01 in order to determine, if there is a significant difference in the solution quality due choosing one energy difference measure over the other.

Data Table 3.3 shows for any two configurations C_1, C_2, consisting of a MAXMINFAIRSA variant and an energy difference measure, the number of instances on which C_1 is better than C_2.

Results According to the data in Table 3.3, ΔE_{ps} shows the worst performance among the three energy difference measures: For each MAXMINFAIRSA variant, ΔE_{ps} is better than the other configurations on 24 instances, which is overall far behind the other energy difference measures. The best performance is achieved using the energy difference measure ΔE_{cw}, which is better than the other configurations on 92 instances (MMFSA-GLBOP) and 79 instances (MMFSA-LSAP). The data also indicates that MMFSA-GLBOP is overall superior to MMFSA-LSAP in our setting: In total, the former is significantly better than the latter 145 times, and the latter is significantly better only 68 times.

Experiment 2: Impact of the MMF-CB-CTT Decomposition

In the following, we will investigate in detail the performance implications of choosing one variant of MAXMINFAIRSA over the other.

Experiment Design We compare the performance of MMFSA-LSAP and MMFSA-GLBOP on the ITC2007 competition instances `comp01`, ..., `comp21` from [DGS]. On each instance we performed 50 independent runs for each variant. In each run, we perform 10^6 SA iterations of the algorithm in order to determine performance differences that occur "on the long run". We use the energy difference measure ΔE_{cw}, which showed the best performance in the first experiment. The quality a MAXMINFAIRSA run is given in terms of the penalty tuple of the best timetable found by the algorithm. For each MAXMINFAIRSA variant, we determined the best and the average allocation over the 50 runs on each instance. The average penalties have been computed using the isomorphism ρ from Theorem 3.12: The penalty tuples were mapped to integers, then the average was calculated and rounded to the nearest integer. Finally, the result was mapped back to the corresponding equivalence class of penalty tuples. We also performed the one-sided WMW test with a significance level of 0.01 [Wil45] to determine, if one of the two algorithm variants yields significantly better results.

Data Table 3.4 shows the best penalty tuples obtained by MMFSA-LSAP and MMFSA-GLBOP on the 21 ITC2007 competition instances. In addition to the best penalty tuples, Table 3.4 shows the average allocations over the 50 runs for each instance. According to the WMW test, MMFSA-GLBOP yields significantly better results than MMFSA-LSAP on instances `comp02`, `comp06`, `comp07`, `comp08`, `comp10`, `comp13`, `comp14`, `comp17`, `comp18`, `comp19`, `comp20` and `comp21`. In contrast, MMFSA-LSAP is not significantly better than MMFSA-GLBOP on any of the 21 instances with this significance level.

Results According to the data in Table 3.4 MMFSA-GLBOP finds better solutions than MMFSA-LSAP on 18 instances, while the best solutions found by MMFSA-LSAP are better on `comp04` and `comp18`. A comparison of the average allocations shows that in this respect, MMFSA-GLBOP outperforms MMFSA-LSAP on 16 instances while it is beaten on the instances `comp05`, `comp08`, `comp15` and `comp21`. The results of the Wilcoxon test are consistent with the data in Table 3.4. We can conclude, that using the decomposition presented in Sect. 3.5 is clearly the better choice, since MMFSA-GLBOP generates better timetables than MMFSA-LSAP when solving the same number of subproblems. Please note however, that if we employ the timeout as it was required for the ITC2007 competition (see [DGMS07]), then MMFSA-LSAP is the better choice, because significantly more iterations are performed within the given timeout.

Experiment 3: Reference Solutions

For reference, we give the best results obtained by the best MAXMINFAIRSA configuration according to the previous experiments. We compare these results with the known best results from the website [DGS].

Table 3.4 Comparison of the best and average objective values of the solutions found by MMFSA-LSAP and MMFSA-GLBOP on the 21 CB-CTT instances from [DGS] for 50 independent runs per instance and per algorithm

	MMFSA-GLBOP		MMFSA-LSAP	
Instance	Best	Average	Best	Average
comp01	**$5^2, 0^{12}$**	**$6, 5, 4^5, 3^4, 2, 1, 0$**	$5^2, 1, 0^{11}$	$6, 5^3, 4, 2^2, 1^5, 0^2$
comp02	**$2^{32}, 1^5, 0^{33}$**	**$4^5, 3^{18}, 2^{16}, 1^{18}, 0^{13}$**	$4, 3^3, 2^{26}, 1^{12}, 0^{28}$	$5^2, 4^{30}, 3^4, 2^{17}, 1^{16}, 0$
comp03	**$6^4, 4^{11}, 2^{23}, 1, 0^{29}$**	**$6^5, 5^{40}, 4^8, 3^2, 2^4, 0^9$**	$6^4, 4^{12}, 2^{27}, 1^3, 0^{22}$	$6^{10}, 5^9, 4, 3^{24}, 2, 1^3, 0^{20}$
comp04	$6^4, 4^2, 2^4, 1^7, 0^{40}$	**$6^4, 4^3, 3^6, 2^{34}, 1^8, 0^2$**	**$6^4, 4^2, 2^4, 0^{47}$**	$6^4, 4^3, 3^{11}, 2^{33}, 1^4, 0^2$
comp05	**$19^2, 18^3, 17^3, 16^5, 15^2, \ldots$**	$20^3, 18^6, 16, 15^9, 14^2, \ldots$	$19^2, 18^3, \ldots, 14^7, \ldots$	**$20^2, 19^4, 18^{11}, 16^3, \ldots$**
comp06	**$12, 4, 2^{29}, 1^{15}, 0^{24}$**	**$12, 5^6, 4^2, 3^9, 2^{43}, 1^4, 0^5$**	$12, 4^3, 3, 2^{31}, 1^{12}, 0^{22}$	$12, 5^{11}, 4^3, 3^5, 2^4, 1^5, 0^{41}$
comp07	**$6, 2^{14}, 1^{18}, 0^{44}$**	**$6, 4, 3^{21}, 2^9, 1^{15}, 0^{30}$**	$6, 3, 2^{30}, 1^{25}, 0^{20}$	$6, 4^2, 3^{22}, 2^7, 1^{11}, 0^{34}$
comp08	**$6^4, 4^2, 2^9, 1^6, 0^{40}$**	$6^4, 4^3, 3^{23}, 2^{15}, 1^{14}, 0^2$	$6^4, 4^2, 2^{11}, 1^4, 0^{40}$	$6^4, 4^4, 3^3, 2^9, 1^{20}, 0^{21}$
comp09	**$6^9, 4^{11}, 2^{18}, 1^2, 0^{35}$**	**$6^{10}, 5^3, 4^8, 3^{21}, 2^{12}, 1^{13}, 0^8$**	$6^9, 4^{14}, 2^{12}, 1^6, 0^{34}$	$6^{10}, 5^5, 4^4, 3^{10}, 2^{10}, 1^{17}, 0^{19}$
comp10	**$2^{12}, 1^5, 0^{50}$**	**$4, 3^{34}, 2^{13}, 0^{19}$**	$2^{20}, 1^9, 0^{38}$	$4^{10}, 3^4, 2^{24}, 1^6, 0^{23}$
comp11	0^{13}	0^{13}	0^{13}	0^{13}
comp12	**$10^4, 8^{26}, 7, 6^{50}, 5^2, \ldots$**	**$12^2, 10^2, 9^{45}, 8^2, 7^{12}, \ldots$**	$10^5, 8^{26}, 6^{48}, 5^6, 4^{42}, \ldots$	$12^3, 11^{19}, 10^9, 9^2, 8^{21}, \ldots$
comp13	**$6^6, 4^4, 2^{12}, 1^2, 0^{42}$**	**$6^6, 4^6, 3^{11}, 2^{24}, 1^6, 0^{13}$**	$6^6, 4^4, 2^{21}, 1^4, 0^{31}$	$6^6, 4^7, 3^{11}, 2, 1^{29}, 0^{12}$
comp14	**$8^4, 4^3, 2^{13}, 1^3, 0^{37}$**	**$8^4, 4^5, 3^5, 2^{32}, 1^{11}, 0^3$**	$8^4, 4^3, 2^{16}, 1^2, 0^{35}$	$8^4, 4^{13}, 3^{22}, 2^{10}, 1^7, 0^4$
comp15	**$6^4, 4^{10}, 2^{22}, 1^4, 0^{28}$**	$6^{11}, 5^3, 4^8, 3^{13}, 2^7, 1^2, 0^{24}$	$6^4, 4^{11}, 3, 2^{25}, 1^5, 0^{22}$	**$6^8, 5^3, 4^{18}, 3, 2^{12}, 1^{18}, 0^8$**
comp16	**$4^5, 2^{14}, 1^3, 0^{49}$**	**$5^5, 4^{10}, 3^6, 2^{19}, 1^6, 0^{25}$**	$4^5, 3, 2^{17}, 1^7, 0^{41}$	$5^{12}, 4^4, 3^{16}, 2^4, 1^{22}, 0^{13}$
comp17	**$10^2, 6^2, 4^5, 2^{31}, 1^8, 0^{22}$**	**$10^2, 6^2, 5, 4^{32}, 2^6, 0^{27}$**	$10^2, 6^2, 4^8, 3, 2^{25}, 1^{12}, 0^{20}$	$10^2, 6^2, 5^9, 4^{29}, 3^{16}, 2^7, 1^5$
comp18	$6, 4^{18}, 3, 2^{15}, 1^4, 0^{13}$	**$7, 6^{19}, 5^6, 4^3, 3^2, 2^9, 1^{11}, 0$**	**$6, 4^{17}, 3^3, 2^{14}, 1^3, 0^{14}$**	$7^4, 5^5, 4^{20}, 3^9, 2^4, 1^9, 0$
comp19	**$6^4, 4^6, 2^{12}, 1^5, 0^{39}$**	**$6^{14}, 5^9, 4^7, 2^{18}, 1^6, 0^{12}$**	$6^4, 4^7, 2^{14}, 1^{10}, 0^{31}$	$7^2, 6^4, 5^{16}, 4^2, 3^{19}, 2^4, 1^{16}, 0^3$
comp20	**$4^3, 3^2, 2^{27}, 1^{16}, 0^{30}$**	**$5^2, 4^{24}, 3^4, 2, 1, 0^{46}$**	$4^5, 3^2, 2^{34}, 1^{13}, 0^{24}$	$5^{16}, 4^6, 3^2, 1^7, 0^{47}$
comp21	**$10, 6^4, 4^{15}, 3, 2^{25}, 1^7, 0^{25}$**	$10, 6^{10}, 5^{33}, 4^4, 2, 1^6, 0^{23}$	$10, 6^4, 5, 4^{16}, 2^{27}, 1^4, 0^{25}$	**$10, 6^9, 5^{15}, 4^6, 3, 2^{31}, 0^{15}$**

The best results are highlighted in bold face

Table 3.5 Penalty tuples of the current best solutions from [DGS] and the best solutions found by MAXMINFAIRSA

Instance	Known best solution	MAXMINFAIRSA ($\Delta E = \Delta E_{cw}$)
comp01	$5^2, 0^{12}$	$5^2, 0^{12}$
comp02	$\mathbf{4, 2^{10}, 0^{59}}$	$4, 2^{29}, 1^3, 0^{37}$
comp03	$13, 10^3, 9, 7^2, 6^4, 5^{13}, 4, 2^6, 0^{37}$	$\mathbf{6^4, 4^{11}, 2^{19}, 1^2, 0^{32}}$
comp04	$7, 6^3, 5^4, 4^2, 2, 0^{46}$	$\mathbf{6^4, 4^2, 2^4, 1^2, 0^{45}}$
comp05	$41^2, 36^7, 35^5, 32^5, 31^6, \dots$	$\mathbf{19^2, 18^3, 17^3, 16^5, 15^3, 14^9}, \dots$
comp06	$12, 7^2, 5^4, 2^3, 0^{60}$	$\mathbf{12, 4^2, 3, 2^{18}, 1^{18}, 0^{30}}$
comp07	$\mathbf{6, 0^{76}}$	$6, 2^{10}, 1^{28}, 0^{38}$
comp08	$7, 6^3, 5^4, 4^2, 2^2, 0^{49}$	$\mathbf{6^4, 4^2, 2^8, 1^3, 0^{44}}$
comp09	$10^5, 9, 7^{10}, 6^6, 5^{10}, 4, 2, 0^{41}$	$\mathbf{6^9, 4^{11}, 2^{15}, 1, 0^{39}}$
comp10	$\mathbf{2^2, 0^{65}}$	$2^{16}, 1^7, 0^{44}$
comp11	0^{13}	0^{13}
comp12	$45, 30^{14}, 28, 27^2, 26^5, 25^{19}, \dots$	$\mathbf{10^4, 8^{24}, 7^4, 6^{41}, 5^5, 4^{33}}, \dots$
comp13	$8, 7, 6^5, 5^7, 4^2, 2^3, 0^{47}$	$\mathbf{6^6, 4^4, 2^{14}, 1^6, 0^{36}}$
comp14	$8^4, 7, 5^2, 2^6, 0^{47}$	$\mathbf{8^4, 4^3, 2^{11}, 1^6, 0^{36}}$
comp15	$10^3, 9^3, 7, 6^4, 5^{13}, 4, 2^7, 0^{36}$	$\mathbf{6^4, 4^7, 2^{23}, 1^3, 0^{31}}$
comp16	$7^2, 5^7, 4, 0^{61}$	$\mathbf{4^3, 2^{13}, 1^9, 0^{46}}$
comp17	$10^2, 6^3, 5^9, 2^4, 0^{52}$	$\mathbf{10^2, 6^2, 4^7, 3, 2^{29}, 1^7, 0^{22}}$
comp18	$17, 15, 14, 13, 11, 10, 9^2, 5^{19}, \dots$	$\mathbf{6, 4^{15}, 2^{13}, 1^4, 0^{19}}$
comp19	$13, 7, 6^4, 5^2, 4, 2^7, 0^{50}$	$\mathbf{6^4, 4^6, 2^{14}, 1^4, 0^{38}}$
comp20	$\mathbf{2^2, 0^{76}}$	$4^2, 2^{26}, 1^{11}, 0^{39}$
comp21	$12, 11, 10^4, 9, 7^4, 6^4, 5^{12}, 4, \dots$	$\mathbf{10, 6^4, 5, 4^{14}, 2^{25}, 1^6, 0^{27}}$
DDS1	$\mathbf{5^{21}, 2^9, 0^{69}}$	$7, 6^5, 5^6, 4^{12}, 3^6, \dots$
DDS2	0^{11}	0^{11}
DDS3	0^9	0^9
DDS4	$\mathbf{15^3, 2, 0^{101}}$	$17^3, 16^3, 15^{14}, 14^{13}, \dots$
DDS5	0^{44}	0^{44}
DDS6	$\mathbf{0^{62}}$	$2^{15}, 1^{12}, 0^{35}$
DDS7	0^{37}	0^{37}
erl.2011-2	$20, 19^2, 17^5, 16, 15^{14}, 14^8, \dots$	$\mathbf{18, 17^7, 16^8, 15^{19}}, \dots$
erl.2012-1	$22, 20^5, 17, 15^4, 12^{10}, 10^{19}, \dots$	$\mathbf{17, 16^9, 15^9, 14^{22}}, \dots$
erl.2012-2	$\mathbf{56, 53^3, 51^{10}, 49^8, 48, 46}, \dots$	$56^9, 54^5, 52^{13}, 50^2, \dots$
erl.2013-1	$19, 17, 15^4, 14^2, 12^6, 11^3, \dots$	$\mathbf{16^2, 15^3, 14^{23}, 13^{13}}, \dots$

Lowest penalties with respect to the max-min fair comparison are highlighted in bold face

Experiment Design On each of the ITC2007 competition, DDS, and Erlangen instances, we perform 50 independent runs of MMFSA-GLBOP with energy difference measure ΔE_{cw}. We report the best penalty tuple obtained for each instance.

Data The data in Table 3.5 shows a comparison of the penalty tuples of the known best solutions from [DGS] with the best penalty tuples found by MMFSA-GLBOP.

Results First of all, for instances `comp11`, `DDS2`, `DDS3`, `DDS5`, and `DDS7`, the penalty tuples of the best existing solutions and the best solution found by MAXMINFAIRSA are identical. For these instances the known best solutions are optimal and have no soft constraint violations. Thus, these solutions are also optimal with respect to max-min fairness. Therefore, MAXMINFAIRSA is able to find optimal solutions to five out of 32 instances. We can also conclude from the data shown in Table 3.5, that the timetables generated by MAXMINFAIRSA have significantly less penalty for the worst-off curricula and the penalty is more evenly distributed across the curricula. The biggest improvements can be observed on the ITC2007 competition instances. This means that although max-min fair timetables may have a higher total penalty according to the CB-CTT objective function, they may be more attractive from the students' perspective, because each student notices an unfortunate arrangement of his/her curriculum. Furthermore, we can observe that if the total penalty of a known best solution is rather low, then it is also good with respect to max-min fairness. For several instances in this category, for example `comp02`, `comp07`, `comp10`, `comp20`, `DDS1`, and `DDS6`, the solution found by MAXMINFAIRSA is not as fair as the known best solution. We can conclude that if there is not much penalty to distribute between the stakeholders, it is not necessary to enforce a fair distribution of penalty.

3.6.3 The Tradeoff Between Fairness and Efficiency

The JFI-CB-CTT problem formulation from Definition 3.9 allows us to investigate the trade-off between fairness and efficiency. Concerning the fairness indices shown in Table 3.2, we can observe that for all known best solution the fairness index is greater than 0.8, with instance `DDS1` being the single exception. In the following, we use the algorithm Archived Multi-Objective Simulated Annealing (AMOSA) [BSMD08] to explore the tradeoff between fairness and efficiency in the JFI-CB-CTT problem model. AMOSA is a variant of SA for multi-objective optimization problems. It maintains an *archive*, which is a set of Pareto-optimal solutions. The algorithm outputs the archive after the minimum temperature has been reached. If the archive size exceeds a given limit during the optimization process, clustering is applied in order to keep a small number of solutions that realize diverse trade-offs. The price of clustering is that the solutions output by the algorithm are not necessarily Pareto-optimal with respect to the set of all solutions encountered by the algorithm. In our setting, clustering was not used, because the number of solutions in the archive was small. In order to determine the acceptance probability, AMOSA needs to measure the amount of Pareto-non-domination, which is done by counting the how many solutions in the archive are dominated by the current solution and vice versa. As before, we use the KX neighborhood exploration.

Experiment Design Since we do not expect from a general multi-objective optimization algorithm to produce solutions as good as the best CB-CTT solvers, we will consider the following scenario to explore the trade-offs between fairness and efficiency: starting from the known best solution from [DGS], we examine how much increase in total penalty is necessary in order to increase the fairness further. We will take as examples the six instances from the ITC2007, whose known best solutions have the highest total penalty, i.e., `comp03`, `comp05`, `comp09`, `comp12`, `comp15` and `comp21`. The temperature levels for AMOSA were set to $\vartheta_{max} = 20$ and $\vartheta_{min} = 0.01$; the cooling rate was set according to Eq. (3.10) with a timeout 192s as determined by the ITC2007 benchmark executable. We will investigate the tradeoff between fairness and efficiency from the set of solutions output by AMOSA.

Data The plots in Fig. 3.3 show the Pareto-optimal solutions from the AMOSA archive. The arrows point to the starting point, i.e., the best available solutions to the corresponding instances with respect to the CB-CTT objective function. The straight lines that go through the initial solutions show a possible trade-off between fairness and efficiency: the slopes were determined such that a 1 % increase in fairness yields a 1 % increase in penalty.

Results For instances `comp05` and `comp21` solutions with a lower total cost than the previously known best solutions were discovered by this approach.[1] For the instances shown in Fig. 3.3, the solutions remain close to the trade-off lines up to a fairness of 94–97 %, while a further increase in fairness demands a significant increase in total cost. In fact, for the instances `comp09` and `comp21`, the fairness index can be increased by 3.5 % and 1.4 %, respectively, without increasing the total penalty at all. For the instances `comp05`, `comp09` and `comp15`, there are several solutions below the trade-off lines. Picking any of the solutions below these lines would result in an increased fairness without an equally large increase in the amount of penalty. This means picking a fairer solution might well be an attractive option in a real-world application. For `comp05` for example, the fairness of the formerly best known solution with a total penalty of 291 can be increased by 5.4 % at 302 total penalty, which is a 3.8 % increase.

In summary, improving the fairness of a timetable that has been optimized for a low overall penalty as a post-processing step seems to be a viable approach for practical decision making. This means we can use proven existing algorithms to generate a good starting point, explore fairer alternatives and then choose the desired trade-off.

[1]Recently, Antony Philips further improved the total penalty by one for `comp21`. His solution matches the known lower bound and is thus optimal, see [DGS].

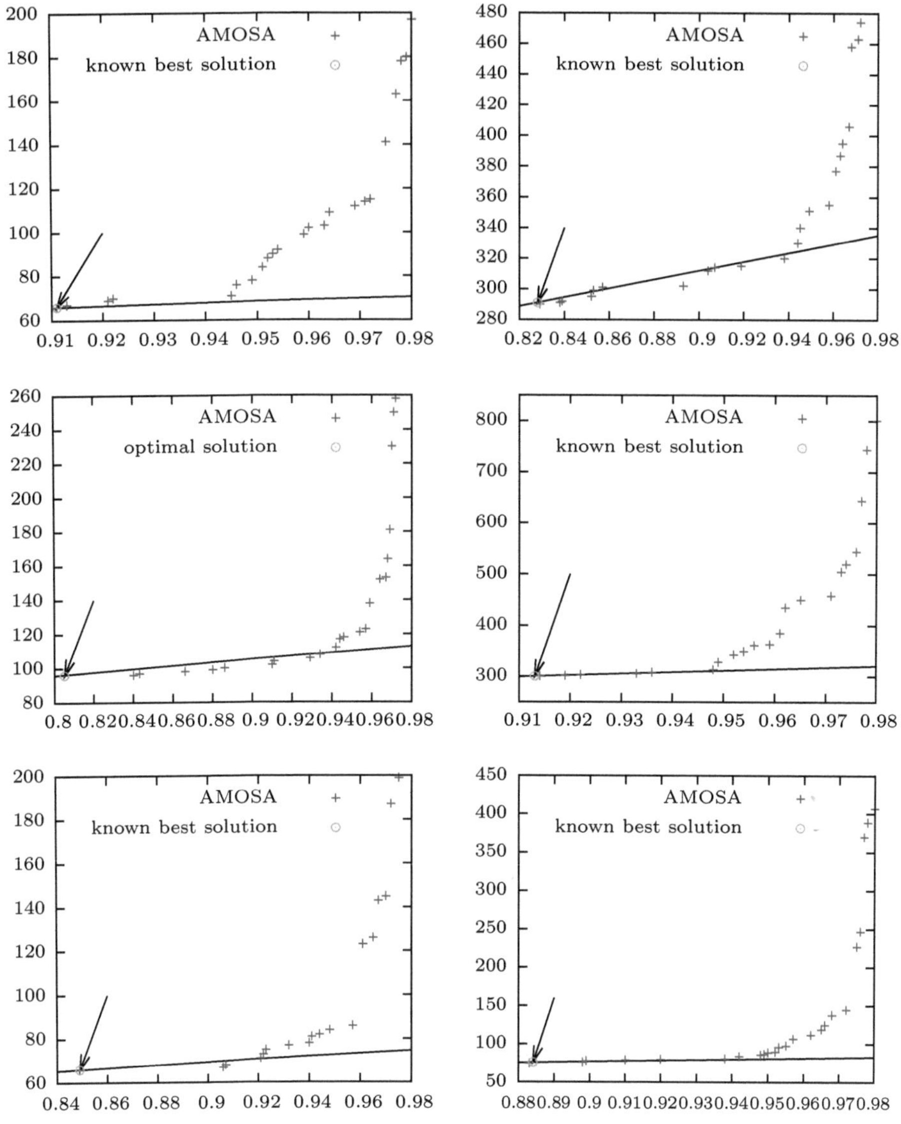

Fig. 3.3 Non-dominated solutions found by the AMOSA algorithm for the JFI-CB-CTT versions of instances `comp03`, `comp05`, `comp09`, `comp12`, `comp15` and `comp21`. All graphs show the fairness index on the horizontal axis and the amount of penalty on the vertical axis

Chapter 4
Real-World Academic Course Timetabling

Although only a few may originate a policy, we are all able to judge it.

Pericles of Athens

The attempt of creating course timetables be means of combinatorial optimization at the school of engineering at the Friedrich-Alexander-Universität Erlangen-Nürnberg (FAU) is a rather recent one. Since the 2011 summer term, the timetables for each semester have been generated using a custom software which has been created alongside with the research presented in this thesis. Course timetables at a university are created for real people, and the computational problems that need to be solved are determined by a social and organizational framework. This framework varies considerably across institutional and legislational borders and imposes unique requirements in each case [McC07, MR12]. Here, we present a case study on automating the creation of course timetables at the FAU school of engineering. We will address the formal problem as well as the organizational framework, feedback mechanisms, acceptance issues, and quality management.

The implementation of the course timetabling process has largely been driven by the following goals:

G1 The implementation has to be finished by the 2011 winter term, within a time-frame of roughly two and a half years.
G2 Establish acceptance among students, lecturers and administration.
G3 Minimize the cost of creating timetables.

The deadline G1 was the result of significant organizational changes expected for the 2011 winter term and the ensuing terms, mostly due to a large increase of the student numbers and a transformation of the structure of the programs of study. A timetable can only be accepted by students, lecturers, and administration, if their various requirements on the schedule are sufficiently respected. Thus, prerequisites of goal G2 are (1) a model that reflects these requirements well and (2) suitable means of quality control to check the validity of the resulting timetables. In the long term, acceptance has been established, if the suggested timetabling process is perceived as the preferred way of creating timetables by the various stakeholders. Creating course timetables is a recurring and tedious task. Basically, any minute

M. Mühlenthaler, *Fairness in Academic Course Timetabling*, Lecture Notes in Economics and Mathematical Systems 678, DOI 10.1007/978-3-319-12799-6_4

spent by lecturers and administration on creating or modifying timetables is a loss, so it is a natural goal to automate the timetabling process as much as possible.

So far, goal G1 has certainly been met. The timetabling system has been in use since 2011 and therefore it is the generally accepted tool for creating the course timetables at the FAU school of engineering. Concerning goal G2, the reception of the course timetables by students and lecturers is discussed in Sect. 4.6. As mentioned above, the key to goal G3 is automation. We will discuss organizational aspects of our course timetabling implementation in Sect. 4.3, with a brief discussion of which parts of the process have been automated and who is involved in the remaining steps.

Theory Versus Practice

Practical implementations of the course timetabling process are typically based on specialized models, which reflect the specific requirements at a particular institution. On the other hand, much of the timetabling research relies on simplified models in order to study and compare the performance of different algorithms. This distinction between theoretical and practical timetabling seems to be widely accepted in the literature [McC07,SH07,MR12,SPMFOB13]. In [SH07], Schimmelpfeng and Helber raise the point that, in the context of automated course timetabling, teacher, student, and administration satisfaction are rarely studied and reported in the literature. Furthermore, McCollum pointed out that it is unclear, in how far improvements on the simplified benchmarking models lead to practical benefits [McC07]. We can conclude that most of the course timetabling research focuses on modeling aspects and computational results while real-world implementation aspects are mentioned rather casually. In our case study, we report on the real-world aspects and provide, for example, empirical data on student satisfaction and lecturer feedback.

4.1 Related Work

We give a brief overview of related case studies on practical implementations of course timetabling systems. Furthermore, we give a brief review of available commercial and non-commercial software systems for creating course timetables at universities.

4.1.1 Related Case Studies

Carter reports on a course timetabling system which has been used successfully between 1985 and 2000 at the University of Waterloo [Car01]. He estimated that the development of the system took 40 person years. In addition to the technical aspects of the system, he gives a sketch of the overall milestones for creating a course timetable for the next semester. The timetabling process starts with a student pre-registration, which provides the students' course selections required to create an initial timetable. The initial timetable is then created and revisions are made by timetabling representatives and students. Decomposition techniques are applied in order to cope with the large size of the instances (17,000–20,000 undergraduate students, 1,400 courses with roughly 3,000 course sections). Carter also points out the political dimension of course timetable creation in order to motivate the addition of availability and preference requirements of the lecturing staff to the model:

> Practical course timetabling is 10% graph theory and 90% politics.

Carter reports improvements with respect to the room utilization as well as the number of satisfied student requests in comparison to manual timetable creation. Furthermore, the tool has been used for an a priori evaluation of the impact of replacing 40 smaller classrooms with 30 generally larger classrooms. Due to maintainability and portability issues, the system has been replaced by a commercial solution after being in use for 15 years.

In [RM03], Rudová and Müller report, among other things, on some practical experiences with a timetabling system used at Purdue university. The timetabling system has been introduced in order to allow for a more efficient room utilization in the face of increasing enrollments. The timetables are created after the students have enrolled in their courses ("post-enrollment"), so exact student demands are known. Due to the large problem size (roughly 29,000 students and 750 classes), the overall problem is decomposed into a central problem dealing with large lectures and several departmental subproblems. Similar to [Car01], they combine the usual course timetabling task with student sectioning. Concerning the acceptance of the automatically generated timetables, they observed that balancing interests between stakeholders is essential:

> Balancing instructor time preferences was found to be a critical factor in developing acceptable solutions since an earlier attempt [Moo91] to automate timetabling at Purdue was unsatisfactory largely because solutions heavily favored instructors who impose the most constraints.

The case study by Schimmelpfeng and Helber [SH07] reports on the implementation of a timetabling system using an Integer Linear Programming (ILP)-based optimization. Their approach has been tailored to the requirements at the School of Management and Economics of the University of Hannover, Germany. To the best of our knowledge, this is the most detailed report on implementation experiences and feedback by the lecturers in the literature. We share most of the experiences and observations reported by Schimmelpfeng and Helber regarding the implementation of the model-based timetabling as well as the timetabling process used prior to the model-based approach. In particular, they also experienced an initial skepticism towards a formalization of the timetabling problem, which can partly be explained by a perceived "loss of individual temporal freedom" [SH07]. There seems to be quite a similar organizational style compared to the School of Engineering at the FAU: Each day of the week is divided into 2 h timeslots, in which lectures of a typical 90 min duration are fitted, such that conflicts and a fixed schedule of external courses (in their case, engineering courses) are respected. They use an elaborate model, which includes pre-assignments, precedence requirements and a diverse set of preference requirements. The instance for the 2006 summer term course schedule includes, among other entities, 99 teachers, 156 courses, and 29 conflict groups. The computational problem was solved using CPLEX [IBM] as well as open-source optimization software. Please note that the problem size (in terms of the number of entities and conflicts) is significantly lower compared to the instances dealt with, for example in [Car01,MMR07] and at the FAU school of engineering. Schimmelpfeng and Helber report positive results on the quality of the timetables and the acceptance of the optimized timetables by the professors and administration. A survey showed that the professors and assistants are satisfied with the planning process as well as the resulting timetables. The case study does not report on the student satisfaction.

4.1.2 Available Software

Managing the assets and activities at a university is a complex task, so it should come as no surprise that there is specialized software available in order to support many administrative tasks. Considering the big picture, creating course timetables is just one—yet central—task among many different administrative activities including also the management of student enrollments, examinations, grades, and lab equipment, just to name a few. Moreover, relevant data about the courses and rooms needs to be accessible by the stakeholders. For this purpose, most timetabling software systems offer some kind of web-based interface which connects to a database backend. In the following, we will give a brief overview over four established timetabling systems. One of them, Unitime [Uni], is open-source, which means that the program source code is publicly available under a permissive license,[1] so the program can be used and modified free of charge. The other three systems are closed-source. All four of the providers offer commercial support for their software and the implementation at particular institutions.

Unitime is an open source timetabling software, which is used for instance at Purdue University for creating course and examination timetables as well as performing student sectioning on a university-wide level for 39,000 students. It offers web-based tools for data management and optimization. At Purdue University, course

[1]GNU General Public License, see http://www.gnu.org/licenses/.

timetables are created in a *post-enrollment* setting. That is, students first choose their courses and then the timetabling system tries to minimize the corresponding event conflicts. The software has been adapted to a curriculum-based model for the Faculty of Education at Masaryk University [MR12]. The optimization in Unitime is performed by a custom constraint solver that showed very good performance in the International Timetabling Competition 2007 (ITC2007) [Mül09]. The performance of the optimizer can be considered state-of-the-art. The Unitime developers also offer a commercial consulting and support service for Unitime users [Uni].

The closed source solutions we consider here are Easystaff EasyAcademy [Eas], Scientia Syllabus Plus Enterprise Foundation [Syl], and Mimosa Scheduling Software [Mim]. Feature-wise, they seem to be quite similar. However, EasyAcademy and Syllabus Plus seem to be more web and service focused than Mimosa. All three systems claim a substantial user base in educational institutions: Scientia claims market leadership with more than 450 users of Enterprise Foundation [Syl, Enterprise Foundation Bruchure], Mimosa lists 300 institutions [Mim], and EasyAcademy is used at more than 30 Italian institutions [Eas]. All of the systems support automated and manual creation of timetables. From publicly available information it is difficult to judge the integration cost of such software systems and whether the software can be customized in a satisfactory manner to the requirements of a particular institution. Therefore, the "make-or-buy" decision is far from trivial.

The use of available (commercial) software for creating course timetables for the FAU school of engineering had been considered prior to the implementation of a custom solution described in this chapter. A decision against an off-the-shelf solution was made, mainly because severe customization of the software would have been required: The desired level of integration with the existing tools (see Sect. 4.3) and the implementation of a custom timetabling problem model that suits the organizational style at the FAU school of engineering (see Sects. 4.4 and 4.5) were expected to be beyond the capabilities of an existing product. Therefore, on the bottom line, the integration cost was identified to be the main cost driver due to support contracts and custom software were required for matching data exchange interfaces between the different software systems.

4.2 Motivation

We will briefly review how course timetables were created at the FAU school of engineering before the software-based approach was introduced in 2011. Next, we will discuss the shortcomings of the manual approach as well as the challenges faced that finally lead to decision that a more systematic approach was needed.

Prior to 2011, the course timetables for the FAU were created by the following simple procedure:

1. Create a copy of last year's course timetable.
2. Apply required changes (i.e., add, remove, or reschedule courses).
3. Resolve clashes.

Steps 2 and 3 were performed in a decentralized fashion: Each lecturer was responsible for the courses he or she offered. So, if a new course was offered or a clash needed to be resolved, adequate rooms and timeslots had to be booked by the corresponding lecturers. All relevant information about course schedules is publicly available in a web-based software called UnivIS [Con], which has been developed at the school of computer science and has been in use for more than two decades. Using UnivIS, it is relatively easy to perform the administrative tasks involved such as looking for alternative timeslots, booking rooms, and many other things. This simple approach is widely accepted, requires minimal effort by a central authority (step 1 is trivial), and has been used successfully university-wide for many years. So why should we change anything?

We will discuss a few rather general issues with the manual approach. First of all, the approach relies on the assumption that course schedules for the next year will be more or less the same as the current ones, so there will likely not be many issues to resolve. If this assumption does no longer hold, and we will argue shortly why it does not, then creating the course schedules becomes quite costly in terms of the time spent by lecturers on fixing the timetable. In particular, fitting a course in an existing timetable without introducing clashes may require other events to be moved, which is hard to coordinate in the decentralized setting. We observed that in order to deal more efficiently with changes in the timetable, policies based on seniority and first-come-first-serve emerged over the years. Such policies discriminate against new courses and, to some extent, discourage cooperative behavior, since the established courses generally occupy attractive rooms and timeslots. Furthermore, due to the changes being made in an incremental and decentralized fashion, it is hard to monitor the overall quality of the timetables from the students' perspective. Very specific knowledge about the structure of the involved programs of study is required in order to avoid introducing new problems by fixing others. In summary, the manual approach is not very robust against changes.

However, rather significant changes were faced by German universities due to a transformation from the established *Diplom* and *Magister Artium* degrees towards bachelor/master degrees in the context of the Bologna process. At the FAU school of engineering, this transformation resulted in a differentiation of the existing study programs as well as a more stringent organization of these programs. Very strict requirements on the timetables result from the detailed specifications of the curricula and the limited time the students have to complete their studies in the new bachelor and masters programs. In addition to these structural changes, a doubling of the number of freshmen was expected for the winter term 2011, because of two cohorts finishing the Bavarian secondary education simultaneously. A further increase of the number of students was expected due to the suspension of compulsory military service. Thus, a number of universities, including the FAU, decided to offer programs beginning in the summer terms in addition to the traditional programs beginning in the winter terms. These rather abrupt changes were accompanied by a general trend towards pursuing higher education, which became manifest in an increasing total number of students the years before 2011 [Reg12].

It was anticipated that these changes cannot be dealt with in a satisfactory manner by the manual scheduling procedure. The two main concerns were an expected room shortage due to the increasing number of students and guaranteeing conflict-freeness between mandatory subjects in all programs of study offered at the school of engineering. A room shortage would have caused additional costs for renting suitable rooms. As a result, funding was granted for a project with the goal of solving the course timetabling problem at the school of engineering within a time-frame of roughly two and a half years, ranging from April 2009 to October 2011. The research presented in this thesis as well as the course timetabling software that is currently used at the school of engineering resulted from this project.

It should be noted that other schools at the FAU have not pursued similar efforts. This is partly because the structural changes and the room shortage were expected to be most severe in the school of engineering. Furthermore, due to the interdependence between study programs and the comparatively strict specification of the individual programs, the school of engineering seemed to be a promising candidate for the use of computer-assisted timetabling. This situation does not generally carry over to the other schools.

The anticipated significant increase in the number of freshmen indeed occurred: Fig. 4.1 shows the number of freshmen enrolled at the FAU school of engineering from the 2006 winter term to the 2013 winter term. Programs beginning in the summer terms were offered from the 2009 summer term (2009-1) and have received significant attention until the 2012 summer term (2012-1). Apart from the expected peak around 2012 winter term, due to the two cohorts finishing their secondary education, we can observe that the number of freshmen has settled around 1,800. This is almost a doubling compared to the number of freshmen in 2006 and 2007.

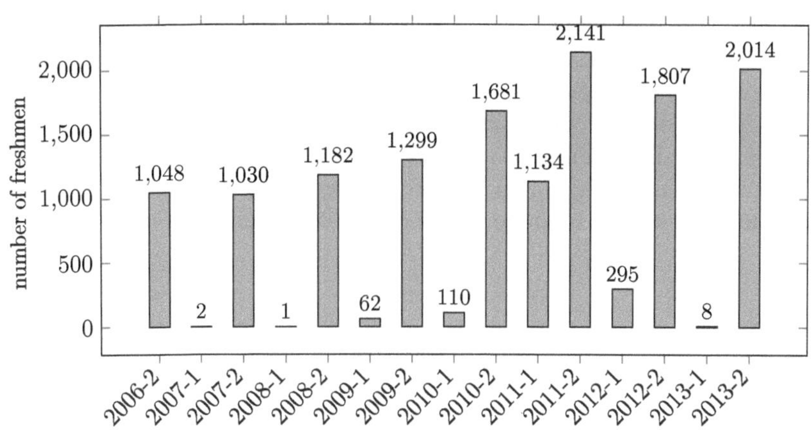

Fig. 4.1 The number of freshmen enrolled at the FAU school of engineering from the 2006 winter term (2006-2) to the 2013 winter term (2013-2)

4.3 The Course Timetabling Process

In the spirit of [Car01] we outline the process of creating the course timetables for the next semester at the FAU school of engineering. There are several conflicting goals and interests that need to be considered in the process. First, the effort spent on creating timetables should be minimal for everyone who is involved in the process since scheduling is not a productive activity in itself. Second, for the teaching staff it is generally desirable to have their timetables available as soon as possible, so they can safely plan their other activities. Third, the course timetable for the school of engineering depends on lectures and rooms from various other schools. The other schools create their timetables manually using the procedure outlined in Sect. 4.2 and typically finalize their timetables just before the beginning of the scheduled semester. Fourth, in order to ensure that the timetables satisfy the requirements of the different stakeholders, the timetable creation is an iterative process. In each iteration, a prospective timetable is checked by a defined group of people, e.g., administration, lecturing staff, and students. Modifications are performed on request such that the resulting timetable is acceptable for everyone. Quality control is a necessity, but it takes time and resources. The overall process is a compromise between the different objectives that has worked reasonably well for all involved parties since the 2011 winter term.

The timetabling process is shown in Fig. 4.2. Each step involves a variety of people and the tasks are performed alongside with their other activities. None of them has a full-time to creating schedules. The overall duration of the timetabling process is roughly 3 months. In the following, we give a brief description of the process:

Data Acquisition Lecturers and administrative staff provide the course and curriculum data in a structured way in UnivIS [Con], a web-based system that stores and visualizes data on courses, rooms, staff, and related things. This system has been in use for roughly two decades at the FAU. Despite being a legacy system, it seemed the best option for the course data acquisition, because of the tight timeframe of the project, interdependencies with other schools, all of which also use UnivIS, and because the staff is familiar with it. Fortunately, only minor modifications were required to adapt it to the requirements of the automated course timetabling. In most

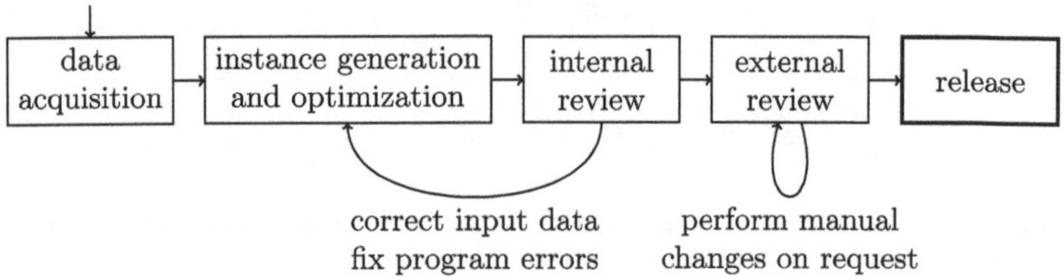

Fig. 4.2 The course timetabling process

cases, the course data can just be copied from the last year, which causes a minimal amount of work. Availability and preference data are collected in another web-based system, which has been developed specifically for this purpose. The data acquisition is usually completed in roughly 10 days.

Instance Generation, Optimization and Internal Review The task of generating instances from course and curriculum data has been fully automated. This task is quite complex and will be discussed separately in Sect. 4.5. After creating a candidate timetable, it is reviewed by a quality manager and representatives of the different study programs. The goal of the internal review is to catch errors caused by incorrect input data. Typical errors are the following:

- An external event needs to be considered by the scheduling software, but it is not.
- An event must only be scheduled in a particular room according to the input data, but the room is too small for the estimated number of students.
- An event cannot be scheduled because no timeslot are available.
- An incomplete curriculum specification leads to missing conflicts.

As a general observation, checking validity within the model is a simple task, while checking the validity of the instance data requires expert knowledge. Thus, parts of the error checking have been automated, but the errors can only be resolved by the experts. Typically, several (less than ten) iterations of the internal review process are required until the instance and the resulting timetable reflect the real-world requirements sufficiently well. This phase of the process is completed within roughly 2 months. It is hard to estimate the actual amount of work in person months, because everyone involved reports the errors in a standard bugtracking software in a rather casual fashion. Between 30 and 90 errors were reported in each semester. There is a clear trend towards fewer reported errors. Manual modifications to the timetable may be needed in order to handle corner cases. The manual modifications are performed once, just before the external review.

External Review During the external review, lecturers and student representatives check the timetables and may request changes. So far, all requested changes were rather minor and could be performed by hand. Performing another iteration of instances generation, optimization and internal review has never been necessary. Typical requests are:

- The room assignment of some event needs to be changed.
- A conflict between two elective or free-choice courses needs to be resolved.

The timeframe of the external review phase is roughly 2 weeks.

Release After the release, the timetable is publicly available in UnivIS. Events that were not considered during the process, typically seminars and lab exercises, are now fitted in the timetable by the lecturers.

4.4 The TF-CB-CTT Problem

The TechFak Curriculum-based Course Timetabling (TF-CB-CTT) problem is the problem model which is used for the recurring task of creating the course timetables at the school of engineering ("Technische Fakultät") of the FAU. This problem model is an abstraction designed to suit the organizational demands placed on the creation of course timetables. We will discuss these requirements during our presentation of the model. First, we attempt to illustrate very briefly the general setting. For each program of study offered at the school of engineering, there is a more or less strictly specified curriculum for each semester, that is, a set of compulsory and elective courses, which are typically taught in the form of lectures, seminars, tutorials, and lab exercises. Each course is taught by one or more lecturers. There are rooms of various sizes for various purposes associated with different organizational units, for example, chairs, departments, and the overall school of engineering. Most rooms of the school of engineering are located at the southern campus in Erlangen ("Südgelände"). There are some curricular interdependencies with other schools, which are situated at the city campuses in Erlangen and Nürnberg.

4.4.1 *Problem Formulation*

The TF-CB-CTT problem is in essence the University Course Timetabling Problem (UCTP) with availability requirements and additional soft constraints. That is, an instance of the search problem consists of timeslot P, events E, rooms R, a conflict graph G and an availability function $\alpha : E \rightarrow \mathcal{P}(P \times R)$ as defined in Sect. 2.1. In order to adapt this basic problem formulation to the requirements at the FAU school of engineering, we extend the basic entities of the model and use as additional entities *courses*, *curricula*, *teachers*, *days*, and *timeslots*, which are based on the Curriculum-based Course Timetabling (CB-CTT) problem model proposed in [DGMS07] and discussed in Sect. 3.1.3. The courses, curricula, and teachers determine in part the event conflicts. However, they can not sufficiently model all event conflicts occurring in our setting, so we keep the conflict graph from the UCTP model and assume that it is in agreement with the conflicts specified by these entities. The additional/extended entities are defined as follows:

- *Courses.* A course is a set of events and has a fixed number of students. In the context of the CB-CTT problem, the events are also referred to as *lectures* [DGMS07]. The set of courses is a partition of the events. Any two lectures of a course are conflicting since they have common students. Each course has one or more teachers. This is an extension of the CB-CTT model and reflects the situation at the FAU school of engineering: for example, per scheduled course there were on average 1.2 teachers in the winter term 2012. Additionally, for each course there is a Boolean value indicating if its lectures should be arranged such that two lectures are in adjacent timeslots.

- *Teachers.* A teacher gives a set of courses and is thus in charge of a set of lectures. Any two lectures given by the same teacher are conflicting.
- *Curricula.* A curriculum is a set of courses. Any two lectures of two courses in a curriculum are conflicting. Note that there is no one-to-one correspondence between curricula in the model and curricula in the programs. Since students typically have a number of choices among tutorials and elective subjects, a curriculum in the TF-CB-CTT model corresponds to a set of courses that must not be taught simultaneously as soon as a student has picked one of the possible combinations.
- *Rooms.* Each room has a capacity, which corresponds to the number of available seats. Additionally, for each room there is a Boolean value indicating if only a limited number of events may be placed in this room.
- *Days* and *Timeslots.* We include in the model a set of weekdays on which there are teaching activities. Each day is divided into a fixed number of *timeslots*. Each timeslot $p \in P$ is identified with a unique combination of a day and a timeslot.

In order to model room and timeslot preferences for individual events, we include the following two preference functions:

- *Period preferences* $\beta : E \times P \to \mathbb{N}$,
- *Room preferences* $\gamma : E \times R \to \mathbb{N}$.

The interpretation of the preference values is the following: For a given event, a room or timeslot with a preference of zero is considered a "best" choice. Higher preference values indicate that a rooms or timeslots are less attractive. Preferences have been specified in a similar fashion for creating course timetables at Purdue University [RM03]. For practical purposes, preferences and availabilities are specified per course and per lecturer at the FAU school of engineering. The preference values of the individual events are set accordingly when a TF-CB-CTT instance is generated.

For clarity of the presentation, we give a formal yet somewhat preliminary definition of the CB-CTT problem here and provide a detailed description of the soft constraints separately in Sect. 4.4.2. The TF-CB-CTT is a combinatorial optimization problem and the task is to find a feasible timetable that minimizes a cost function. Similar to the CB-CTT problem, the cost function is a weighted sum of soft the constraint violations.

Definition 4.1 (TechFak Curriculum-based Course Timetabling (TF-CB-CTT) Problem)

INSTANCE: A TF-CB-CTT instance consists of entities, an availability function and a preference function as specified above. A timetable τ is an assignment $\tau : E \to P \times R$. A timetable is called feasible, if it is clash-free, overlap-free and satisfies all availability requirements. For a TF-CB-CTT instance $\mathcal{I}$, let $\mathcal{S}(\mathcal{I})$ be the set of feasible timetables.

TASK: Given a TF-CB-CTT instance $\mathcal{I}$, find a timetable $\tau^* \in \mathcal{S}(\mathcal{I})$ which minimizes the objective function $f : \mathcal{S}(\mathcal{I}) \to \mathbb{N}$ specified in Sect. 4.4.2.

4.4.2 Soft Constraints

The TF-CB-CTT model includes ten different soft constraints, which address workload, distribution and preference requirements of lecturers and students. We include in the model the four soft constraints S1–S4 from the CB-CTT problem formulation, that is, the RoomCapactiy, MinWorkingDays, IsolatedLectures, and RoomStability requirements, as described in Sect. 3.1.3. Additionally, we include the following requirements:

S5 (Student Load) For each curriculum, the number of daily lectures must be within a given range. Each lecture below the minimum and above the maximum of the range counts as one violation. This constraint occurs in the UD3–UD5 variants of the CB-CTT problem proposed in [BDCDGS12].

S6 (Double Lectures) We use a slight variation of the double lectures constraint from [DGMS07], which stipulates, that two lectures of a course that occur on the same day must be grouped. That is, they must be scheduled in adjacent timeslots and in the same room. In the TF-CB-CTT model, a grouping of the lectures of a course can also be requested. However, all lectures of a course need to be grouped, not only if they occur on the same day. If grouping is requested for a course, then any occurrence of a non-grouped lecture of this course counts as one violation.

S7 (Period Preferences) Assigning to an event e a timeslot p of attractiveness $\beta(e, p) = k$ results in k violations.

S8 (Room Preferences) Assigning to a lecture e a room r of attractiveness $\gamma(e, r) = k$ results in k violations.

S9 (Teacher Load) For each teacher, there is a maximum number of lectures per day. Each lecture exceeding this maximum number counts as one violation.

S10 (Restricted Room Usage) If indicated for a given room, the automatically scheduled events may occupy at most $h \in \mathbb{N}$ timeslots. Each additional occupied timeslot in such a room counts as one violation. This constraint prevents rooms that are used for staff meetings and other administrative purposes from being filled up with automatically scheduled events. In contrast to the room availability requirements, this constraint does not exclude specific timeslots for a given room.

The TF-CB-CTT objective function f measures the quality of a timetable $\tau \in \mathcal{S}(\mathcal{I})$ by taking the weighted sum of the number of violations of the soft constraints S1–S10:

$$f(\tau) = \sum_{1 \le i \le 10} w_{\mathrm{S}i} f_{\mathrm{S}i}(\tau) \ , \tag{4.1}$$

where $w_{\mathrm{S1}}, \ldots, w_{\mathrm{S10}} \in \mathbb{N}$ and $f_{\mathrm{S1}}, \ldots, f_{\mathrm{S10}} : \mathcal{S} \to \mathbb{N}$ functions counting the number of violations of the soft constraints S1–S9, respectively. Currently, we set

$$(w_{\mathrm{S1}}, \ldots, w_{\mathrm{S10}}) = (1000, 7, 1, 1, 2, 800, 7, 5, 10, 5) \ .$$

The large weight of the constraint S1 (Minimum Working Days) reflects how undesirable timetables are, for students and lecturers alike, if the number of lectures per day exceeds what course has been designed for. The situation is similar for constraint S6 (Double Lectures). These two constraints should be satisfied unless there is really no other possibility, so they are essentially treated like hard constraints in comparison to the other soft constraints.

The UD5 formulation of the CB-CTT from [DGMS07] considers the travel distance between two consecutive lectures to ensure that students have sufficient time to move to the location of the next lecture. Since the FAU has various campuses and faculties spread all over the cities Erlangen and Nürnberg, there are sometimes considerable travel times between the lectures. The current solution is to have a 30 min break between every two 90 min lecture slots. This break can be used for traveling. Including travel times might be a beneficial addition to the TF-CB-CTT model in the future. However, there are many practical issues that are implicitly taken care of by the current solution. In particular, it is not necessary to deal with the huge number of different combinations of subjects of different faculties that potentially have implications for the travel costs. Also, the required amount of communication between different faculties is minimal. Certainly, if other faculties were to participate in the automatic timetabling procedure considering travel costs would become necessary.

4.5 Instance Generation

Having defined a process and a computational model for the course timetabling problem at the FAU school of engineering, it may seem like the remaining challenge is largely to solve TF-CB-CTT instances. From a computational-complexity point of view, this seems reasonable. In practice however, generating instances from the available data is a very complex step with huge implications for the perceived quality of the resulting timetable. Here, we are not referring to the quality of a timetable within the model, but rather to the validity of the instances and how well the timetable is accepted by students and lecturers. Any error that happens during the instance generation, any piece of data that is missing can have catastrophic consequences and render the resulting timetable completely useless for practical purposes, no matter how little the soft constraints are violated in the model. We will briefly describe several selected aspects of how TF-CB-CTT instances are generated from the available data in our setting and which issues arise besides ensuring sufficient input data quality. The first two of these issues, determining the course conflicts and estimating the number of students attending a course, are specific to the curriculum-based course timetabling model and would not occur, for example, if each student were required to enroll in his or her courses of choice prior to the timetabling process ("post-enrollment" course timetabling).

A central structure of the UCTP and the TF-CB-CTT alike is the conflict graph which represents event conflicts. In our practical timetabling application, the conflict graph has to be constructed from the courses that are offered in a given semester and a description of the curricula. A curriculum is given by a semester and a program of study, for instance Electrical Engineering, semester three. At the FAU school of engineering, there are three main categories of subjects: compulsory subjects, elective subjects and free choices. There are slight variations in how these terms are used and interpreted across different programs. For our purposes, a subject which is compulsory for a certain curriculum has to be attended by every student enrolled in this curriculum. The compulsory subjects in a curriculum are conflicting, and the elective subjects must not be taught simultaneously with the compulsory subjects. Thus, each student is able to attend the mandatory classes. Due to the typically large number of elective subjects it is generally impossible to guarantee conflict-freeness for all sets of elective subjects for the available number of timeslots. For example, in the 2013 winter term, 15 curricula had at least 30 elective subjects, and the maximum number of elective subjects over all curricula was 57. In this situation, maximizing the students' choices among the elective subjects by minimizing conflicts between them seems desirable at first sight. However, there are more and less popular combinations of elective subjects and heuristic information is required to actually give the students the choices they want. This is why we decided to mark certain combinations of subjects as conflicting based on feedback from the lecturers. The free choice subjects are too diverse to be considered in the conflict graph. Our general impression is that the conflict graph is indeed a useful abstraction, but determining if two events are conflicting or not is far from trivial.

Since the exact course choices of the students are unknown during the timetabling, the number of students who will be attending a given course needs to be estimated. The estimation is based on the number of students enrolled in the various curricula related to the course. Exact data on the number of enrolled students is available as soon as students are in their second semester. The number of freshmen is estimated based on data from the ministry of education and historical data, with an added safety margin. If a course is elective for a given curriculum, only a certain unknown fraction of the students in the curriculum will attend it. Furthermore, an elective course may be offered in several semesters of the same program, which further complicates the situation. Our approach is rather simplistic: We assume that per study program a fixed fraction d of the students in the largest curriculum chooses an elective subject. In our implementation, we set $d = 0.1$. This value is a result of our experience and is likely very specific to the FAU school of engineering. For courses with multiple sections—mostly tutorials—the students are evenly distributed among the individual sessions.

Another delicate subtask of the instance generation consists of determining the room preferences. Each course has as associated administrative entities a *chair*, a *department* (Electrical, Electronic and Communication Engineering, Chemical and Biological Engineering, Materials Science and Engineering, Mechanical Engineering, Computer Science) and a school. Depending on these entities, different rooms are preferable for a course. For each course, the available rooms are divided into seven preference classes, given in decreasing order of preference:

1. Rooms that were selected as "favorite" for the course or rooms that are managed by the chair.
2. Rooms managed by the associated department.
3. Rooms managed by the school.
4. Rooms managed by another chair of the same department.
5. Rooms managed by another department of the same school.
6. Rooms managed by the (close by) chemistry department.
7. All other rooms.

A room in preference class i receives the preference value $i - 1$. Thus, assigning an event to one of its "favorite" rooms causes no violation of soft constraint S8.

In contrast to the room preferences, the timeslot preferences and lecturer availabilities are collected in a straightforward fashion from a web-based tool referred to as the *Timeslot Manager*. There are only two preference values indicating if a timeslot is *preferred* or *neutral*. Assigning a course to a preferred timeslot results in the preference value 0, while neutral timeslots yield the preference value 1. Clearly, the availability requirements place quite severe constraints on the search space. Murray, Müller and Rudová observed the "tendency of the solution to favor those who place the most restrictions" [MMR07]. Therefore, cooperative behavior is discouraged by the timetabling system in this respect and some intervention is required. In order to gain the acceptance of the lecturers we tried to place minimal restrictions on the specification of the availability requirements. Our approach is the following: The timeslots in the evening (18:00 to 20:00) and in the morning (from 8:00 to 10:00) were identified as the most unpopular timeslots. Since teaching activity in these timeslots is necessary, each teacher is required to be available at least once a week in the morning and also once a week in the evening. There are no further restrictions concerning the availability of lecturers.

In [MMR07,MM10], student sectioning is combined with the course timetabling task. In our application, some rudimentary student sectioning for tutorial classes is performed manually by restricting tutorials sessions to subsets of the involved curricula. The sectioning reduces the overall conflicts of the tutorial sessions and thus typically results in a better placement of the tutorials with respect to the timetables of the individual curricula. Due to the lack of information on the students' actual course choices, we try to avoid a too restrictive sectioning and to retain

Table 4.1 Statistical data on the Erlangen instances

Instance	|C|	|E|	|R|	|U|	Co	CA	RA	RS	RO	CS
Erlangen-2011-2	755	827	176	1,917	0.02	0.68	0.82	0.34	0.19	4.67
Erlangen-2012-1	764	829	110	3,435	0.03	0.73	0.81	0.69	0.31	4.13
Erlangen-2012-2	850	930	132	3,678	0.03	0.69	0.80	0.59	0.29	4.98
Erlangen-2013-1	738	825	137	3,249	0.03	0.68	0.77	0.59	0.26	4.44
Erlangen-2013-2	705	788	140	3,497	0.04	0.71	0.77	0.61	0.24	4.26

some flexibility in choosing tutorial sessions. We suspect that due to the incomplete information, adding student sectioning to the model will not yield a practical benefit.

In the following, we present some statistical data on the actual TF-CB-CTT instances generated for the FAU school of engineering ranging from the 2011 winter term to the 2013 winter term. Please note, that the instance generation as well as the model have been refined over time. The first four columns of Table 4.1 show the number of courses $|C|$, the number of event $|E|$, the number of rooms $|R|$, and the number $|U|$ of curricula. The remaining entries measure the *conflict density* (Co), the *course availability* (CA), the *room availability* (RA), the *room suitability* (RS), the *room occupation* (RO), and the *curriculum size* (CS), which are defined as follows:

- Co: The number of edges of the conflict graph divided by the number of edges of the complete graph on the events.
- CA: The number of available timeslots per course divided by the number of course-timeslot pairs.
- RA: The number of available timeslots per room divided by the number of room-timeslot pairs.
- RS: The number of suitable rooms per course divided by the number of course-room pairs.
- RO: The number of events divided by the number of available timeslots in all rooms, irrespective of the size of the rooms.
- CS: The average number of courses per curriculum.

Comparable statistical data on CB-CTT instances from the University of Udine used in the ITC2007 is available in [BDCDGS12]. All instance data is available on the website [DGS]. In order to fit in the standard CB-CTT benchmark model, the additional information such as room preferences was removed.

Each instance has 5 teaching days divided into 6 time periods of 2 h each. Therefore, 30 timeslots are available per week. When the automated timetabling system was introduced, we considered adding timeslots on Saturday. Teaching on Saturday would certainly have been very unpopular with staff and students alike. Fortunately, feasible timetables were found having the usual 30 timeslots available. As shown in Table 4.1, the number of rooms is fluctuating significantly between different terms. The reasons include an evolving infrastructure (new buildings, older ones being temporary closed for maintenance, rooms being converted to computer

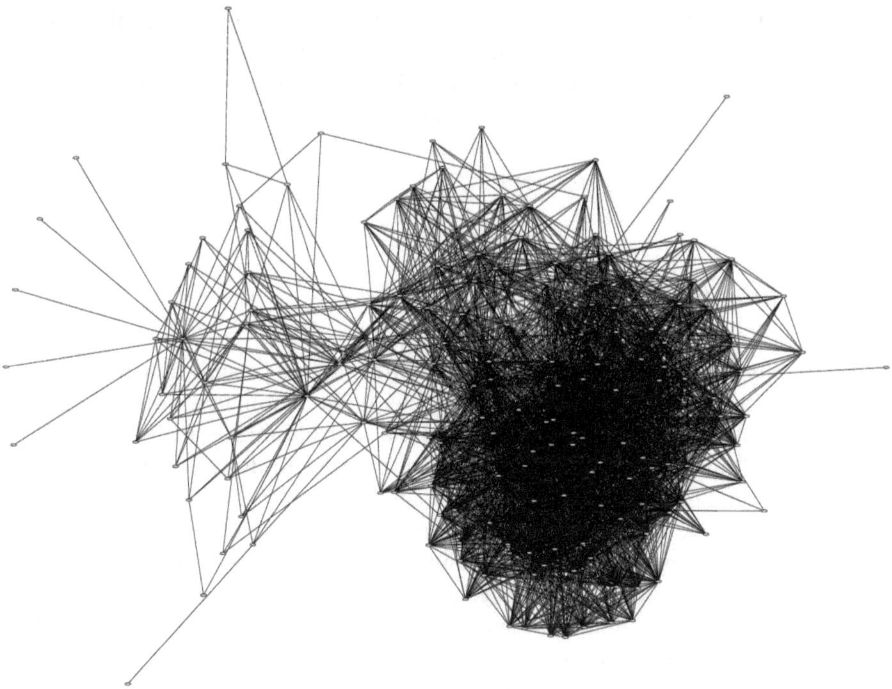

Fig. 4.3 Course conflicts between the electrical engineering courses in the 2013 winter term

labs, etc.), a varying availability of external rooms, and different approaches to selecting from all rooms in the UnivIS database the ones suitable for teaching.

Figure 4.3 shows as an illustration of the conflicts between all courses attended by electrical engineering students in the 2013 winter term. In total there are 180 courses and 2,995 conflicts. The lower part essentially consists of the mandatory courses and their associated tutorials.

4.6 Reception and Feedback

Recall our overall goal of establishing acceptance of the automatically generated timetables among students, lecturers and administration (goal G2). In order to meet this goal, we try to create course timetables in accordance with the stakeholder's preferences. Some of these preferences can be stated explicitly and a-priori, for example by specifying timeslot preferences in the Timeslot Manager. In order to learn about how well a timetable respects the explicit and implicit preferences of the stakeholders, several forms of feedback are included in the timetabling process, see Sect. 4.3. Another feedback mechanism, which has not been discussed so far is the student satisfaction survey, which has been established in order to evaluate and improve the teaching at the FAU. Since the trial run of the automated timetabling system in the 2011 summer term, the students are able to give feedback on the quality of their timetables within the survey. The corresponding part of the survey timetable quality has been designed by the quality management representative. Here, we present and discuss statistical data from the student satisfaction and the Timeslot Manager usage by lecturers.

4.6.1 Students

We are mainly interested in which parts of the timetabling process can be improved according to the students' opinions. Our focus is the automated timetable generation, in particular, model, instance generation and optimization. It is important to note some limitations of using survey data in order to draw conclusions about automated timetabling. First of all, the automated timetabling process only *manages* the course timetable creation given fixed requirements, which are determined by the curriculum design, lecturers' availabilities, arrangements with other schools, and so on. Therefore, it is a delicate task to decide, if features of a timetable have to be attributed to the model and the optimization or the circumstances. For the stakeholders, there is no difference. Second, it should be noted that students generally lack information about the formalisms and how the timetable creation works and therefore we can only draw weak conclusions from their statements. For example, the clash-freeness of a timetable has a very precise meaning in our model. It is guaranteed that for each curriculum all mandatory courses are clash-free. However, the question if a particular student's mandatory courses are clash-free seems to interpreted as "was I able to make the choices I wanted?" according to the comments in the survey. Such preferences are not known during the timetable creation, which is an inherent drawback of the curriculum-based approach. Nevertheless, we present real measured data on the perceived quality of the course timetables and which features of a timetable are considered as unfortunate by the students, based on a rather large number of samples. Additionally, we briefly discuss the comments given in text form by the students.

Each student at the FAU school of engineering may evaluate the courses he or she attended and the quality of the timetable at the end of a term. First, the students evaluate positive features of their individual timetables by giving grades from 1 to 5, where 1 means that the timetable has a certain feature and 5 means the feature is not present. Then they give a binary indication if specific undesirable features are present in their timetables. Additionally, they can give comments as free text. The positive were roughly stated as follows:

- There are no clashes of my mandatory courses.
- I am satisfied with my timetable.
- In comparison to the previous term my timetable has improved.
- There are available seats in my courses.

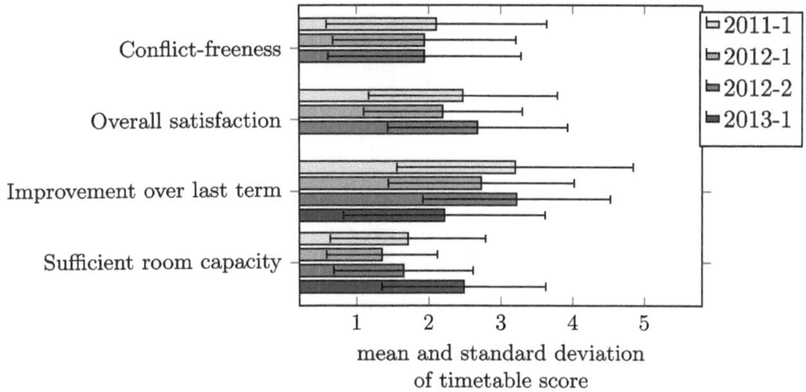

Fig. 4.4 General student feedback on the timetable quality. Scores range from 1 to 5, lower scores are better

Figure 4.4 shows the mean and the standard deviation from the evaluations in the 2011 and 2013 summer terms and the 2012 winter terms. The 2012 winter term data is based on a separate survey, the *FAU Panel 2013*, which contained the same questions on the timetable quality as the other surveys. The number of samples for the individual items ranges from 705 to 950 (2011-1), 799 to 821 (2013-1), and 749 to 1,055 (2012-2). No specific conclusions about a possible improvement of the automated timetable creation can be drawn from the rating of the overall satisfaction and the improvement compared to the previous term. However, the data indicates that sufficient priority is given to respecting the room capacity constraint S2. The clash-freeness feature shows a considerable difference between the timetable quality measured in the model and the perceived quality. Each timetable released to the students and lecturers is guaranteed to be feasible (within the model). Thus, each student should be able to attend all mandatory courses. Unfortunately, none of the students elaborated on the clashes in the free text comment section of the survey.

Figure 4.5 shows the student feedback on six specific undesirable features of the timetables. For each feature, the students indicated if it is present in their individual timetable or not, which gives an impression of how often these features occur. In addition, students could indicate if they are satisfied with their timetable or not. The data shows that the course load per day ("Too many courses per day") is of some concern, but free days, which are possible only at the cost of a high course load on other days, are very desirable ("Lectures spread over too many days"). Moreover, attending courses in the evening seems to be slightly more unpopular than courses in the morning. Although the data in Fig. 4.5 does not indicate that the travel time is a problem for many students, the free text comments contained quite a number of complaints, as we will see shortly. The number of idle gaps in the timetables seem to be the most serious issue. By increasing the weight of the "isolated lectures" soft constraint S4, the number of idle gaps may be reduced. However, gaps may be

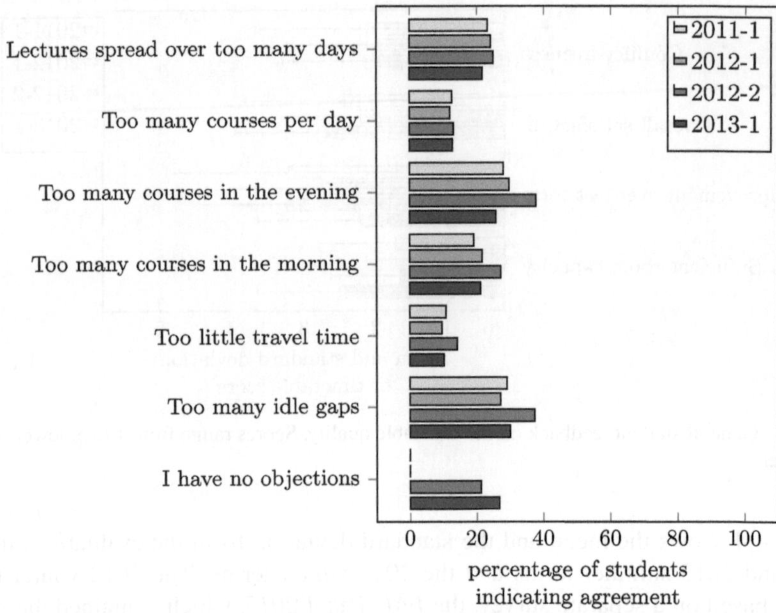

Fig. 4.5 Student feedback on five specific timetable properties and overall satisfaction. Students indicated, if their individual timetable has the corresponding property or not. We give the percentage of YES answers

produced, for example, by a particular choice of tutorial sessions, so more detailed feedback would be desirable, in order to avoid a trial-and-error procedure.

We also evaluated the free text comments on the timetable quality from the FAU Panel 2013 survey on the 2012 winter term timetable, which includes the most samples of all available surveys. From the comments we identified a number of discomforts experienced by the students, which are related to their timetables. As an estimation of the overall severity of the discomforts, we counted how often they were mentioned in the comments. A summary of the results is shown in Table 4.2.

In total, 126 out of 178 comments refer to the discomforts shown in the table above. The remaining comments mostly address organizational issues which are not related to the automated timetable construction. There is a significant number of comments on clashes among elective and free choice subjects. The existence of such conflicts is unavoidable: during the 2012 winter term there were 20 curricula with at least 25 elective subjects, not including free choice subjects. The maximum number of elective subjects of a curriculum was 58. Clash-freeness of the elective subjects thus entails gaps in the individual timetables and/or teaching at unpopular times. Currently, elective subjects can be marked as conflicting on request by the lecturers. As a possible improvement of this approach, a pre-selection of elective subjects by the students could be considered in order to determine the students' preferences. The nuisances of traveling were addressed more often in the free text comments than the

Table 4.2 A summary of how often the various types of discomforts were mentioned in the free text comments of the FAU Panel 2013 student survey

Discomfort	Comments
Lectures on Friday	6
Course load is too high	5
Course load is too low	2
Elective/free choice courses are clashing	30
Too much traveling	30
Lectures late in the evening	14
Lectures early in the morning	4
Lunch break too short	4
Too many gaps	10
Insufficient room size	6

data in Fig. 4.5 suggested. This indicates, that there is a rather serious problem, which concerns only a particular group of students. The school of engineering travel times are not part of the model, but have been considered in a CB-CTT benchmarking variant [BDCDGS12]. The main problem in our application is that lectures at other schools are assumed as fixed and a better coordination between the schools is required in order to minimize the traveling. The data also indicates that lectures and tutorials in the evening timeslots are unpopular. Surprisingly, the idle timeslots in the timetable were not mentioned as often in the text comments as in the explicit review of undesirable features occurring in the timetable.

4.6.2 Lecturers

In contrast to the students, there is no survey available on the satisfaction of the lecturers with the timetabling system. Feedback is given actively by the lecturers via email during the data acquisition and the external review phase of the timetabling process. Most of this feedback is concerned with room changes or moving events away from evening or morning timeslots. Typically, the request in the review phase involve a trade-off between subjective improvement and objective deterioration of the timetable with respect to the model. In order to establish acceptance of the timetabling system among the lecturers, we try to satisfy the lecturers' requests whenever possible and sensible. Figure 4.6 shows the number of requests by lecturers during the data acquisition phase and the external review phase. According to the data, the requests during the data acquisition phase are increasing, while the requests during the review phase are decreasing. Thus, we can conclude that there is a trend towards incorporating requirements in the model, rather than requesting changes after the optimization has been performed.

We observed that scheduling a lecture in an off-campus room often results in a complaint by the lecturer. The off-campus rooms are very unpopular because they involve additional traveling for students and lecturers. Nevertheless, each of

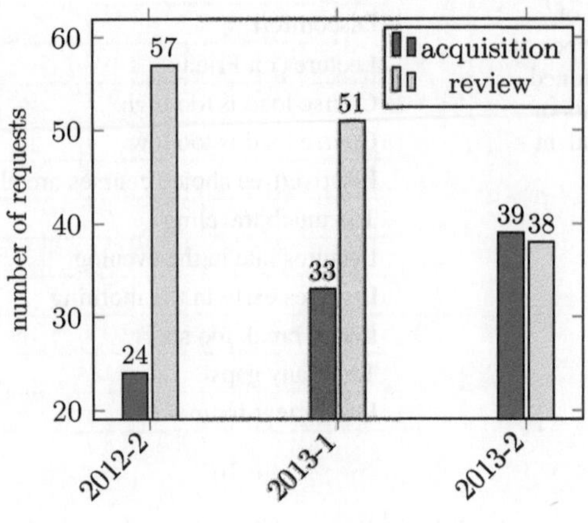

Fig. 4.6 Number of requests by lecturers during the data acquisition phase and the external review phase

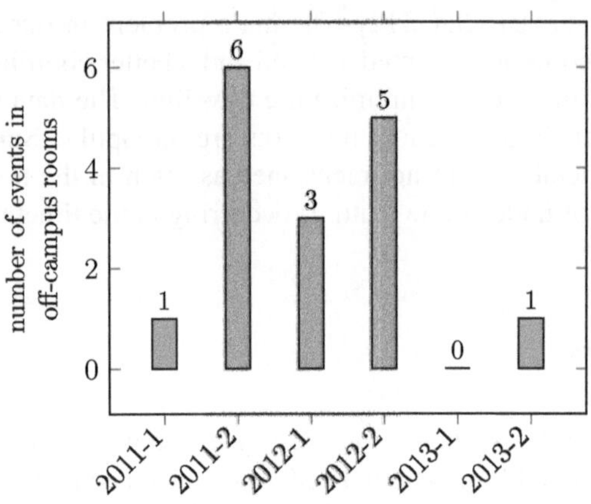

Fig. 4.7 Number of events scheduled in the unpopular off-campus rooms

the off-campus rooms can accommodate at least 400 students and thus, using these off-campus rooms was necessary in order to avoid drastic over-booking. We tried to balance the assignment of the off-campus rooms by ensuring that lecturers are not scheduled in off-campus rooms several terms in a row. As shown in Fig. 4.7, the number of events in off-campus rooms has been decreasing, as shown in Fig. 4.7, which results in fewer complaints. The reasons are twofold: First, the number of freshmen is decreasing slightly again compared to the peak around the 2012 winter term, as indicated in Fig. 4.1. Second, some of the lectures attended by a large number of students have been split, for example, by using video transmission, in consultation with the lecturers, so the lectures could be scheduled on-campus.

Appendix A

A.1 Degeneracy of UCTP Conflict Graphs

In this section, we document the degeneracy of conflict graphs for various University Course Timetabling Problem (UCTP) benchmark sets that do not have availability requirements. This is supplementary data for Section 2.3.2. The tables A.1 and A.2 show degeneracy values of UCTP instance conflict graphs which were computed by repeatedly removing vertices of minimal degree (see Sect. 2.3.1). Entries in bold face indicate that the clash-free timetables are connected with respect to the Kempe-exchange operation due to Thm. 2.10.

A.2 A Meta-heuristics Toolbox

Since course timetabling problems are typically computationally hard, practitioners and researches typically resort to meta-heuristics as solution approaches. In this section, we briefly review several basic meta-heuristic techniques for solving search and optimization problems. This section serves as a supplement to our discussion of different solution approaches to timetabling problems in the previous chapters.

Many search and optimization problems are hard, that is, no algorithms are known that find optimal solutions in polynomial time. Meta-heuristics are a popular class of approaches that attempt to find *good* solutions within reasonable time. They are often the last resort to problems (e.g., course timetabling problems), which cannot be dealt with satisfactorily by other approaches. On the downside, not much has been proved concerning useful properties of the solutions generated by meta-heuristics. Their performance is typically determined experimentally. A large number of meta-heuristics exists that are inspired by how nature solves optimization/adaptation tasks, for example Evolutionary Algorithm [BFM97] and particle swarm optimization (PSO) [Ken10], and ant colony optimization (ACO) [DS04].

M. Mühlenthaler, *Fairness in Academic Course Timetabling*, Lecture Notes
in Economics and Mathematical Systems 678, DOI 10.1007/978-3-319-12799-6

Table A.1 Degeneracy values of the conflict graphs

Instance	deg(G)	Instance	deg(G)	Instance	deg(G)
small_1	54	med_1	59	big_1	60
small_2	**41**	med_2	67	big_2	68
small_3	98	med_3	67	big_3	64
small_4	69	med_4	69	big_4	80
small_5	84	med_5	87	big_5	75
small_6	**24**	med_6	101	big_6	93
small_7	68	med_7	120	big_7	111
small_8	84	med_8	98	big_8	82
small_9	124	med_9	121	big_9	77
small_10	136	med_10	64	big_10	77
small_11	**34**	med_11	97	big_11	76
small_12	**22**	med_12	78	big_12	76
small_13	146	med_13	105	big_13	84
small_14	100	med_14	92	big_14	74
small_15	79	med_15	101	big_15	127
small_16	118	med_16	145	big_16	115
small_17	120	med_17	126	big_17	184
small_18	60	med_18	188	big_18	131
small_19	141	med_19	173	big_19	159
small_20	**28**	med_20	153	big_20	144

The instances are the CB-CTT, PE-CTT and Erlangen benchmarking instances from [DGS]. For each instance there are 45 timeslots. The degeneracy values are marked in bold face if the connectedness of clash-free timetables is established by Theorem 2.10

Table A.2 Degeneracy values of the conflict graphs of the metaheuristic network instances

Instance	deg(G)	Instance	deg(G)	Instance	deg(G)
easy01	**15**	medium01	49	hard01	68
easy02	**19**	medium02	53	hard02	67
easy03	**13**	medium03	52		
easy04	**12**	medium04	51		
easy05	**20**	medium05	47		

For each instance there are 45 timeslots. Degeneracy values are marked in bold face if the connectedness of clash-free timetables is established by Theorem 2.10

Most of the nature-inspired approaches operate on a population of candidate solutions to a problem instance. Another class of metaheuristics is based on the Local Search (LS) paradigm, which aims for an iterative improvement of a single candidate solution [PS98, p. 454]. Examples of LS variants are Simulated Annealing (SA) [KGV83], Tabu Search (TS) [GL97], and Iterated Local Search (ILS) [LMS03]. Without loss of generality, we will talk about minimization problems (Tables A.1 and A.2).

A.2.1 Local Search

In order to find an optimal solution to a problem instance $\mathcal{I}$, an LS algorithm traverses the search space of a problem instance $\mathcal{I}$ using some neighborhood N on the solutions:

$$\mathrm{N} : \mathcal{S}(\mathcal{I}) \rightarrow \mathcal{P}(\mathcal{S}(\mathcal{I})) \ .$$

Algorithm 6 shows the basic LS procedure as given in [PS98, p. 454]. The function `step` selects a neighbor from the neighborhood N:

$$\texttt{step}(t) = \begin{cases} \text{some } s \in \mathrm{N}(t) \text{ s.t. } f(s) < f(t) & \text{if it exists} \\ \text{“no”} & \text{otherwise} \end{cases} ,$$

where f is the objective function. Stochastic Local Search (SLS) algorithms introduce randomness during the initialization and/or the neighborhood selection [HS04, pp. 30–31]. Furthermore, neighbors may be selected even if they are not strict improvements over the current solution in order to escape local optima. This class of algorithms includes for example SA, TS, and ILS.

Algorithm 6: LOCALSEARCH

input : $\mathcal{I}$: problem instance
output: A solution to $\mathcal{I}$

$s \longleftarrow$ initial solution
while `step`$(s) \neq$ “no” **do**
  $s \longleftarrow$ `step`(s)
return s

Simulated Annealing

Simulated Annealing (SA) is a SLS optimization algorithm inspired by the minimization of free energy during the annealing process in metallurgy [KGV83]. An overview of different variants of the algorithm and theoretical results is given in [HJJ03a]. The selection of the initial solution is specific to the problem to be solved. For example, the initialization step for solving the MMF-CB-CTT problem may consist of solving the underlying search problem, see [MW12*]. In each iteration, the SA algorithm selects a neighbor s of the current solution t at random from $\mathrm{N}(t)$. The neighbor s replaces t (“s is accepted”), if it is better. If it is worse then it is accepted with a certain probability that tends to decrease over time. More specifically, the probability of accepting a worsening depends on (i) the current temperature ϑ and (ii) the *energy difference* $\Delta E(t, s)$ between the current solution t and the neighbor s. The accepted solution is determined as follows:

$$\texttt{accept}(t, s) = \begin{cases} s & \text{if } f(s) < f(t) \text{ or } \mathrm{u}(0, 1) \leq \exp(-\Delta E(t, s)/\vartheta) \\ t & \text{otherwise} \end{cases} ,$$

where u(0, 1) draws a number from the interval [0, 1] uniformly at random. The energy difference is typically set to $\Delta E(t, s) = f(t) - f(s)$. In summary, several things happen during each iteration: First, a neighbor s is picked from N(t) at random. Then, t is updated to $\texttt{accept}(s, t)$. Finally, the temperature ϑ is decreased according to the cooling schedule.

A wealth of cooling schedules is available in the literature, see for example [HJJ03b]. A rather generic cooling schedule is *geometric cooling*. It has been applied for example for solving timetabling problems in [TD98, Kos05, MW12*]. Given an initial temperature ϑ_{max}, the cooling schedule decreases the temperature exponentially:

$$\vartheta(i) = \beta^i \cdot \vartheta_{max} ,$$

where i is the iteration.

Tabu Search

Tabu Search (TS) is a SLS variant which utilizes a memory structure, the *tabu list*, in order to guide the search process. We follow the general description of TS given in [Glo89]. The neighborhood exploration is based on so-called *moves*, which transform a solution into a neighbor solution. The tabu list contains *forbidden* and thus renders a part of the neighborhood "tabu". In each iteration, a set S of solutions is sampled from the neighborhood, excluding neighbors resulting from tabu moves. Then the current solution is updated to one of the solutions in S, for example one which minimizes the objective function over S. Then the tabu list is updated, for example, such that the tabu list contains the inverse moves of the last k moves performed, for some $k \in \mathbb{N}$. The *aspiration criterion* may override the tabu status of a neighbor. For instance, if a tabu neighbor improves on the best solution found so far it may still be considered.

A.2.2 Evolutionary Algorithms

Evolutionary Algorithms (EAs) is a class of algorithm inspired by the evolution of species. For an overview of theory and applications of evolutionary algorithms see [BFM97,MF04]. An EA typically maintains a population of candidate solutions

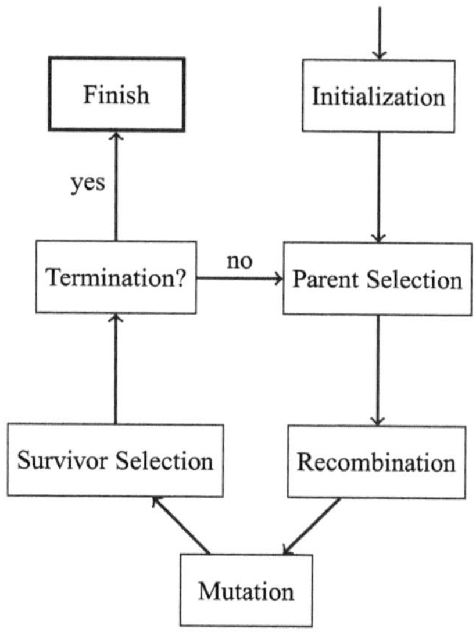

Fig. A.1 The evolutionary cycle, adapted from [Wei02, p. 43]

to a problem instance. The search space is explored by "evolving" the population over time, that is, new candidate solutions are created by mechanisms such *mutation* or *crossover*. Selection mechanisms keep good candidates with respect to the objective function and remove others from the population.

Figure A.1 shows a very general and high-level view of evolutionary algorithms. The initialization step creates at random a population of candidate solutions. Then a selection of promising solutions for recombination is performed. The recombination creates new solutions ("offsprings") by combining features of two or more parent solutions. The offsprings are mutated, that is, modified at random. Then the survivor selection determines from the parents and the offsprings a set of solutions that forms the population in the next iteration. This cycle is repeated until the termination criterion is satisfied. Typically, the algorithm terminates when sufficiently good solutions have been found or a maximum number of iterations has been performed.

Bibliography

[AAN12] Achá, R.A., Nieuwenhuis, R.: Curriculum-based course timetabling with SAT and MaxSAT. Ann. Oper. Res. 1–21 (2012). doi:10.1007/s10479-012-1081-x

[AANORC09] Achá, R.A., Nieuwenhuis, R., Oliveras, A., Rodríguez-Carbonell, E.: Cardinality networks and their applications. In: Kullmann, O. (ed.) 12th International Conference on Theory and Applications of Satisfiability Testing, SAT'09. Lecture Notes in Computer Science, vol. 5584, pp. 167–180. Springer, Berlin (2009)

[AL05] Arntzen, H., Løkketangen, A.: A tabu search heuristic for a university timetabling problem. In: Ibaraki, T., Nonobe, K., Yagiura, M. (eds.) Metaheuristics: Progress as Real Problem Solvers. Operations Research/Computer Science Interfaces Series, vol. 32, pp. 65–85. Springer, New York (2005). doi:10.1007/0-387-25383-1_3

[BB12] Burke, E.K., Bykov, Y.: The late acceptance hill-climbing heuristic. Technical Report, CSM-192, Computing Science and Mathematics, University of Stirling (2012)

[BC96] Baker, B.S., Coffman, Jr., E.G.: Mutual exclusion scheduling. Theor. Comput. Sci. **162**(2), 225–243 (1996). doi:10.1016/0304-3975(96)00031-X

[BC09] Bonsma, P., Cereceda, L.: Finding paths between graph colourings: PSPACE-completeness and superpolynomial distances. Theor. Comput. Sci. **410**, 5215–5226 (2009). doi:10.1016/j.tcs.2009.08.023

[BDCDGS12] Bonutti, A., De Cesco, F., Di Gaspero, L., Schaerf, A.: Benchmarking curriculum-based course timetabling: formulations, data formats, instances, validation, and results. Ann. Oper. Res. **194**(1), 59–70 (2012). doi:10.1007/s10479-010-0707-0

[BDM09] Burkard, R.E., Dell'Amico, M., Martello, S.: Assignment Problems. Society for Industrial and Applied Mathematics, Philadelphia (2009)

[BdWK03] Burke, E.K., de Werra, D., Kingston, J.H.: Applications to timetabling. In: Gross, J.L., Yellen, J. (eds.) Handbook of Graph Theory, 1st edn., Chap. 5.6, pp. 475–483. CRC Press, Boca Raton (2003)

[BEM$^+$10] Burke, E.K., Eckersley, A.J., McCollum, B., Petrovic, S., Qu, R.: Hybrid variable neighbourhood approaches to university exam timetabling. Eur. J. Oper. Res. **206**(1), 46–53 (2010). doi:10.1016/j.ejor.2010.01.044

[BFM97] Bäck, T., Fogel, D.B., Michalewicz, Z. (eds.): Handbook of Evolutionary Computation, 1st edn. IOP Publishing, Bristol (1997)

[BFT11] Bertsimas, D., Farias, V.F., Trichakis, N.: The price of fairness. Oper. Res. **59**(1), 17–31 (2011). doi:10.1287/opre.1100.0865

M. Mühlenthaler, *Fairness in Academic Course Timetabling*, Lecture Notes in Economics and Mathematical Systems 678, DOI 10.1007/978-3-319-12799-6

[BGK+95] Barr, R.S., Golden, B.L., Kelly, J.P., Resende, M.G.C., Stewart, Jr., W.R.: Designing and reporting on computational experiments with heuristic methods. J. Heuristics **1**(1), 9–32 (1995)

[Bir] Birattari, M.: The Race Package. http://cran.r-project.org/web/packages/race (2013). Accessed Sept 2013

[BJL+11] Bonamy, M., Johnson, M., Lignos, I., Patel, V., Paulusma, D.: On the diameter of reconfiguration graphs for vertex colourings. Electron. Notes Discrete Math. **38**(1), 161–166 (2011). doi:10.1016/j.endm.2011.09.028

[BJL+12] Bonamy, M., Johnson, M., Lignos, I., Patel, V., Paulusma, D.: Reconfiguration graphs for vertex colourings of chordal and chordal bipartite graphs. J. Comb. Optim. 1–12 (2012). doi:10.1007/s10878-012-9490-y

[BLS00] Bergmann, R., Ludbrook, J., Spooren, W.P.J.M.: Different outcomes of the Wilcoxon-Mann-Whitney test from different statistics packages. Am. Stat. **54**(1), 72–77 (2000)

[BLSS05] Burke, E.K., Silva, J.D.L., Soubeiga, E.: Multi-objective hyper-heuristic approaches for space allocation and timetabling. In: Ibaraki, T., Nonobe, K., Yagiura, M. (eds.) Metaheuristics: Progress as Real Problem Solvers. Operations Research/Computer Science Interfaces Series, vol. 32, pp. 129–158. Springer, New York (2005). doi:10.1007/0-387-25383-1_6

[BMM+07] Burke, E.K., McCollum, B., Meisels, A., Petrovic, S., Qu, R.: A graph-based hyper-heuristic for educational timetabling problems. Eur. J. Oper. Res. **176**(1), 177–192 (2007)

[BN99] Burke, E.K., Newall, J.P.: A multistage evolutionary algorithm for the timetable problem. IEEE Trans. Evol. Comput. **3**(1), 63–74 (1999). doi:10.1109/4235.752921

[Bon12] Bonsma, P.: The complexity of rerouting shortest paths. In: Rovan, B., Sassone, V., Widmayer, P. (eds.) Mathematical Foundations of Computer Science 2012. Lecture Notes in Computer Science, vol. 7464, pp. 222–233. Springer, Berlin/Heidelberg (2012). doi:10.1007/978-3-642-32589-2_22

[BR91] Burkard, R.E., Rendl, F.: Lexicographic bottleneck problems. Oper. Res. Lett. **10**, 303–308 (1991). doi:10.1016/0167-6377(91)90018-K

[BSMD08] Bandyopadhyay, S., Saha, S., Maulik, U., Deb, K.: A simulated annealing-based multiobjective optimization algorithm: AMOSA. IEEE Trans. Evol. Comput. **12**, 269–283 (2008). doi:10.1109/TEVC.2007.900837

[Bul98] Bullnheimer, B.: An examination scheduling model to maximize students' study time. In: Proceedings of the 2nd International Conference on the Practice and Theory of Automated Timetabling (PATAT), pp. 78–91 (1998). doi:10.1007/BFb0055882

[BYBS10] Birattari, M., Yuan, Z., Balaprakash, P., Stützle, T.: F-race and iterated f-race: an overview. In: Bartz-Beielstein, T., Chiarandini, M., Paquete, L., Preuss, M. (eds.) Experimental Methods for the Analysis of Optimization Algorithms, pp. 311–336. Springer, Berlin/Heidelberg (2010). doi:10.1007/978-3-642-02538-9_13

[BZ80] Burkard, R.E., Zimmermann, U.: Weakly admissible transformations for solving algebraic assignment and transportation problems. Math. Progam. Study **12**, 1–18 (1980). doi:10.1007/BFb0120884

[BZ03] Batagelj, V., Zaversnik, M.: An $O(m)$ algorithm for cores decomposition of networks. CoRR cs.DS/0310049 (2003)

[Car01] Carter, M.W.: A comprehensive course timetabling and student scheduling system at the university of Waterloo. In: Selected Papers from the Third International Conference on Practice and Theory of Automated Timetabling III, PATAT '00, pp. 64–84. Springer, London (2001)

[CDGGS12] Chiarandini, M., Gaspero, L.D., Gualandi, S., Schaerf, A.: The balanced academic curriculum problem revisited. J. Heuristics **18**(1), 119–148 (2012)

[CDGS12] Ceschia, S., Gaspero, L.D., Schaerf, A.: Design, engineering, and experimental analysis of a simulated annealing approach to the post-enrolment course timetabling problem. Comput. Oper. Res. **39**(7), 1615–1624 (2012). doi:10.1016/j.cor.2011.09.014

[CdMRLS11] Constantino, A.A., de Melo, E.L., Romao, W., Landa-Silva, D.: A heuristic algorithm for nurse scheduling with balanced preference satisfaction. In: Proceedings of IEEE Symposium on Computational Intelligence in Scheduling (CISched), pp. 39–45 (2011). doi:10.1109/SCIS.2011.5976549

[Che10] Chen, J.: A new SAT encoding of the at-most-one constraint. In: Proceedings of Constraint Modelling and Reformulation (2010)

[Chi89] Chiu, R., Jain, D.-M.: Analysis of the increase and decrease algorithms for congestion avoidance in computer networks. Comput. Netw. ISDN Syst. **17**, 1–14 (1989)

[CK95] Cooper, T.B., Kingston, J.H.: The complexity of timetable construction problems. In: Practice and Theory of Automated Timetabling, pp. 283–295 (1995). doi:10.1007/3-540-61794-9_66

[CL96] Carter, M.W., Laporte, G.: Recent developments in practical examination timetabling. In: Burke, E.K., Ross, P. (eds.) Practice and Theory of Automated Timetabling. Lecture Notes in Computer Science, vol. 1153, pp. 1–21. Springer, Berlin/Heidelberg (1996). doi:10.1007/3-540-61794-9_49

[CL07] Cioppa, T.M., Lucas, T.W.: Efficient nearly orthogonal and space-filling latin hypercubes. Technometrics **49**(1), 45–55 (2007)

[CLL96] Carter, M.W., Laporte, G., Lee, S.Y.: Examination timetabling: algorithmic strategies and applications. J. Oper. Res. Soc. **47**(3), 373–383 (1996)

[CM01] Castro, C., Manzano, S.: Variable and value ordering when solving balanced academic curriculum problems. In: 6th Workshop of the ERCIM WG on Constraints (2001)

[Con] Config Informationstechnik eG. UnivIS—ein www-basiertes Informationssystem für Hochschulen. http://univis.de/ (2013). Accessed Sept 2013

[Con06] Conover, W.J.: Practical Nonparametric Statistics. Wiley Series in Probability and Statistics, 3rd edn. Wiley, New York (2006)

[Cor05] Corneil, D.G.: Lexicographic breadth first search—a survey. In: Hromkovic, J., Nagl, M., Westfechtel, B. (eds.) Graph-Theoretic Concepts in Computer Science. Lecture Notes in Computer Science, vol. 3353, pp. 1–19. Springer, Berlin/Heidelberg (2005). doi:10.1007/978-3-540-30559-0_1

[CvdHJ08] Cereceda, L., van den Heuvel, J., Johnson, M.: Connectedness of the graph of vertex-colourings. Discrete Math. **308**(5–6), 913–919 (2008). doi:10.1016/j.disc.2007.07.028

[CvdHJ11] Cereceda, L., van den Heuvel, J., Johnson, M.: Finding paths between 3-colorings. J. Graph Theory **67**(1), 69–82 (2011). doi:10.1002/jgt.20514

[DCPT99] Della Croce, F., Paschos, V.T., Tsoukias, A.: An improved general procedure for lexicographic bottleneck problems. Oper. Res. Lett. **24**(4), 187–194 (1999). doi:10.1016/S0167-6377(99)00013-9

[DGMS07] Di Gaspero, L., McCollum, B., Schaerf, A.: The second international timetabling competition (ITC-2007): curriculum-based course timetabling (Track 3). In: Proceedings of the 1st International Workshop on Scheduling, a Scheduling Competition (SSC) (2007)

[DGS] Di Gaspero, L., Schaerf, A.: Curriculum-based course timetabling web-site. http://satt.diegm.uniud.it/ctt/ (2013). Accessed Sept 2013

[DGS01] Di Gaspero, L., Schaerf, A.: Tabu search techniques for examination timetabling. In: Burke, E., Erben, W. (eds.) Practice and Theory of Automated Timetabling III. Lecture Notes in Computer Science, vol. 2079, pp. 104–117. Springer, Berlin/Heidelberg (2001)

[DGS06] Di Gaspero, L., Schaerf, A.: Neighborhood portfolio approach for local search applied to timetabling problems. J. Math. Model. Algorithm **5**(1), 65–89 (2006). doi:10.1007/s10852-005-9032-z

[Die12] Diestel, R.: Graph Theory. Graduate Texts in Mathematics, vol. 173, 4th edn. Springer, Heidelberg (2012)

[DLL62] Davis, M., Logemann, G., Loveland, D.: A machine program for theorem-proving. Commun. ACM **5**(7), 394–397 (1962). doi:10.1145/368273.368557

[DPS13] Dostert, M., Politz, A., Schmitz, H.: Algorithms and complexity of student sectioning for existing timetables. In: Proceedings of the 6th Multidisciplinary International Conference on Scheduling: Theory and Applications (MISTA), pp. 218–229 (2013)

[DS04] Dorigo, M., Stützle, T.: Ant Colony Optimization. MIT Press, Cambridge (2004)

[Due93] Dueck, G.: New optimization heuristics: the great deluge algorithm and the record-to-record travel. J. Comput. Phys. **104**(1), 86–92 (1993). doi:10.1006/jcph.1993.1010

[DVY03] De Veciana, G., Yang, X.: Fairness, incentives and performance in peer-to-peer networks. Seeds **250**(300), 350 (2003)

[dW85] de Werra, D.: An introduction to timetabling. Eur. J. Oper. Res. **19**(2), 151–162 (1985)

[dW97] de Werra, D.: Restricted coloring models for timetabling. Discrete Math. **165–166**(15), 161–170 (1997). doi:10.1016/S0012-365X(96)00208-7

[Eas] EasyStaff EasyAcademy. http://www.easystaff.it (2014). Accessed Jan 2014

[EF70] Edmonds, J., Fulkerson, D.R.: Bottleneck extrema. J. Combin. Theory **8**(3), 299–306 (1970). doi:10.1016/S0021-9800(70)80083-7

[EIS76] Even, S., Itai, A., Shamir, A.: On the complexity of timetable and multicommodity flow problems. SIAM J. Comput. **5**(4), 691–703 (1976). doi:10.1137/0205048

[Erb01] Erben, W.: A grouping genetic algorithm for graph colouring and exam timetabling. In: Burke, E.K., Erben, W. (eds.) Practice and Theory of Automated Timetabling III. Lecture Notes in Computer Science, vol. 2079, pp. 132–156. Springer, Berlin/Heidelberg (2001). doi:10.1007/3-540-44629-X_9

[ES04] Eén, N., Sörensson, N.: An extensible SAT-solver. In: Giunchiglia, E., Tacchella, A. (eds.) Theory and Applications of Satisfiability Testing. Lecture Notes in Computer Science, vol. 2919, pp. 502–518. Springer, Berlin/Heidelberg (2004). doi:10.1007/978-3-540-24605-3_37

[Fal94] Falkenauer, E.: A new representation and operators for genetic algorithms applied to grouping problems. Evol. Comput. **2**(2), 123–144 (1994)

[Fal98] Falkenauer, E.: Genetic Algorithms and Grouping Problems. Wiley, New York (1998)

[FG10] Frisch, A.M., Giannaros, P.A.: SAT encodings of the at-most-*k* constraint. In: Proceedings of International Workshop on Modelling and Reformulating Constraint Satisfaction Problems (2010)

[Fri37] Friedman, M.: The use of ranks to avoid the assumption of normality implicit in the analysis of variance. J. Am. Stat. Assoc. **32**(200), 675–701 (1937). doi:10.1080/01621459.1937.10503522

[FS06] Feldman, A.M., Serrano, R.: Welfare Economics and Social Choice Theory, 2nd edn. Springer, New York (2006). doi:10.1007/0-387-29368-X

[Gin21] Gini, C.: Measurement of inequality of incomes. Econ. J. **31**(121), 124–126 (1921). doi:10.2307/2223319

[GJ79] Garey, M.R., Johnson, D.S.: Computers and Intractability: A Guide to the Theory of NP-Completeness. W.H. Freeman, New York (1979)

[GKMP09] Gopalan, P., Kolaitis, P.G., Maneva, E., Papadimitriou, C.H.: The connectivity of boolean satisfiability: computational and structural dichotomies. SIAM J. Comput. **38**(6), 2330–2355 (2009). doi:10.1137/07070440X

[GKR12] Gorski, J., Klamroth, K., Ruzika, S.: Generalized multiple objective bottleneck problems. Oper. Res. Lett. **40**(4), 276–281 (2012)

[GL97] Glover, F., Laguna, M.: Tabu Search. Kluwer Academic Publishers, Norwell (1997)

[Glo89] Glover, F.: Tabu search: part I. ORSA J. Comput **1**(3), 190–206 (1989)

[GT89] Gabow, H.N., Tarjan, R.E.: Faster scaling algorithms for network problems. SIAM J. Comput. **18**(5), 1013–1036 (1989). doi:10.1137/0218069

[HD05] Hearn, R.A., Demaine, E.D.: PSPACE-completeness of sliding-block puzzles and other problems through the nondeterministic constraint logic model of computation. Theor. Comput. Sci. **343**(1–2), 72–96 (2005). doi:10.1016/j.tcs.2005.05.008

[HJJ03a] Henderson, D., Jacobson, S.H., Johnson, A.W.: The theory and practice of simulated annealing. In: Handbook of Metaheuristics, pp. 287–319. Springer, New York (2003)

[HJJ03b] Henderson, D., Jacobson, S.H., Johnson, A.W.: The theory and practice of simulated annealing. In: Glover, F., Kochenberger, G.A. (eds.) Handbook of Metaheuristics. International Series in OperationsResearch & Management Science, vol. 57, pp. 287–319. Springer, New York (2003). doi:10.1007/0-306-48056-5_10

[HKW02] Hnich, B., Kiziltan, Z., Walsh, T.: Modelling a balanced academic curriculum problem. In: Proceedings of CP-AI-OR-2002, pp. 121–131 (2002)

[HS04] Hoos, H., Sttzle, T.: Stochastic Local Search: Foundations & Applications. Morgan Kaufmann Publishers, San Francisco (2004)

[HvdWZ06] Hothorn, T., Hornik, K., van de Wiel, M.A., Zeileis, A.: A lego system for conditional inference. Am. Stat. **60**(3), 257–263 (2006)

[IBM] IBM ILOG. IBM ILOG CPLEX Optimizer: High-performance mathematical programming solver for linear programming, mixed integer programming, and quadratic programming. http://www.ilog.com/products/cplex/ (2013). Accessed Nov 2013

[IDH+11] Ito, T., Demaine, E.D., Harvey, N.J.A., Papadimitriou, C.H., Sideri, M., Uehara, R., Uno, Y.: On the complexity of reconfiguration problems. Theor. Comput. Sci. **412**(12–14), 1054–1065 (2011)

[JCH84] Jain, R.K., Chiu, D.-M.W., Hawe, W.R.: A quantitative measure of fairness and discrimination for resource allocation in shared computer systems. Technical Report, DEC-TR-301, Digital Equipment Corporation (1984)

[Jer95] Jerrum, M.: A very simple algorithm for estimating the number of k-colorings of a low-degree graph. Random Struct. Algorithm **7**(2), 157–165 (1995). doi:10.1002/rsa.3240070205

[Joh02] Johnson, D.S.: A theoretician's guide to the experimental analysis of algorithms. In: Data Structures, Near Neighbor Searches, and Methodology: Fifth and Sixth DIMACS Implementation Challenges, vol. 59, pp. 215–250. American Mathematical Society, Providence (2002)

[KAJ94] Koulamas, C., Antony, S.R., Jaen, R.: A survey of simulated annealing applications to operations research problems. Omega **22**(1), 41–56 (1994). doi:10.1016/0305-0483(94)90006-X

[Ken10] Kennedy, J.: Particle swarm optimization. In: Encyclopedia of Machine Learning, pp. 760–766. Springer, New York (2010)

[KGV83] Kirkpatrick, S., Gelatt, C.D., Vecchi, M.P.: Optimization by simulated annealing. Science **220**(4598), 671–680 (1983)

[KK06] Kumar, A., Kleinberg, J.M.: Fairness measures for resource allocation. SIAM J. Comput. **36**(3), 657–680 (2006). doi:10.1137/S0097539703434966

[KMT98] Kelly, F.P., Maulloo, A.K., Tan, D.K.H.: Rate control for communication networks: shadow prices, proportional fairness and stability. J. Oper. Res. Soc. **49**(3), 237–252 (1998)

[Kőn16] Kőnig, D.: Über Graphen und ihre Anwendung auf Determinantentheorie und Mengenlehre. Math. Ann. **77**(4), 453–465 (1916). doi:10.1007/BF01456961

[Kos05] Kostuch, P.: The university course timetabling problem with a three-phase approach. In: Burke, E.K., Trick, M. (eds.) Practice and Theory of Automated Timetabling. Lecture Notes in Computer Science, vol. 3616, pp. 109–125. Springer, Berlin/Heidelberg (2005). doi:10.1007/11593577_7

[KRT01] Kleinberg, J.M., Rabani, Y., Tardos, É.: Fairness in routing and load balancing. J. Comput. Syst. Sci. **63**(1), 2–20 (2001). doi:10.1006/jcss.2001.1752

[KS60] Kapur, J.N., Saxena, H.C.: Mathematical Statistics. S. Chand and Company, New Delhi (1960)

[LA87] van Laarhoven, P.J.M., Aarts, E.H.L.: Simulated Annealing: Theory and Applications. Kluwer Academic Publishers, Boston (1987)

[LC10] Lan, T., Chiang, M.: An axiomatic theory of fairness in network resource allocation. www.princeton.edu/~chiangm/fairness.pdf (2010). Extended version of [LKCS10]

[Lew06] Lewis, R.: Metaheuristics for university course timetabling. Ph.D. thesis, Napier University, Edinburgh (2006)

[Lew08] Lewis, R.: A survey of metaheuristic-based techniques for university timetabling problems. OR Spectrum **30**(1), 167–190 (2008). doi:10.1007/s00291-007-0097-0

[LH10] Lü, Z., Hao, J.-K.: Adaptive tabu search for course timetabling. Eur. J. Oper. Res. **200**(1), 235–244 (2010). doi:10.1016/j.ejor.2008.12.007

[LIDLSB11] López-Ibáñez, M., Dubois-Lacoste, J., Stützle, T., Birattari, M.: The irace package, iterated race for automatic algorithm configuration. Technical Report, TR/IRIDIA/2011-004, IRIDIA, Université Libre de Bruxelles (2011)

[LKCS10] Lan, T., Kao, D., Chiang, M., Sabharwal, A.: An axiomatic theory of fairness in network resource allocation. In: Proceedings of IEEE INFOCOM, pp. 1343–1351 (2010). doi:10.1109/INFCOM.2010.5461911

[LL10] Lach, G., Lübbecke, M.E.: Curriculum based course timetabling: new solutions to Udine benchmark instances. Ann. Oper. Res. 1–18 (2010). doi:10.1007/s10479-010-0700-7

[LL12] Lach, G., Lübbecke, M.E.: Curriculum based course timetabling: new solutions to Udine benchmark instances. Ann. Oper. Res. **194**, 255–272 (2012). doi:10.1007/s10479-010-0700-7

[LMS03] Lourenço, H.R., Martin, O.C., Stützle, T.: Iterated local search. In: Glover, F., Kochenberger, G.A. (eds.) Handbook of Metaheuristics. International Series in Operations Research & Management Science, vol. 57, pp. 320–353. Springer, New York (2003). doi:10.1007/0-306-48056-5_11

[LPa] Lewis, R., Paechter, B.: GGA experimental results. http://www.soc.napier.ac.uk/~benp/centre/timetabling/experimentalresults2.htm (2013). Accessed Sept 2013

[LPb] Lewis, R., Paechter, B.: New "harder" instances for the university course timetabling problem. http://www.soc.napier.ac.uk/~benp/centre/timetabling/harderinstances.htm (2013). Accessed Sept 2013

[LP04] Lewis, R., Paechter, B.: New crossover operators for timetabling with evolutionary algorithms. In: Proceedings of the 5th International Conference on Recent Advances in Soft Computing (RASC), pp. 189–195 (2004)

[LP07] Lewis, R., Paechter, B.: Finding feasible timetables using group-based operators. IEEE Trans. Evol. Comput. **11**(3), 397–413 (2007)

[Luc09] Lucarelli, G.: Scheduling in computer and communication systems and generalized graph coloring problems. Ph.D. thesis, Athens University Economics and Business (AUEB) (2009)

[LVM81] Las Vergnas, M., Meyniel, H.: Kempe classes and the Hadwiger conjecture. J. Combin. Theory Ser. B **31**(1), 95–104 (1981). doi:10.1016/S0095-8956(81)80014-7

[LZC11] Liu, Y., Zhang, D., Chin, F.Y.L.: A clique-based algorithm for constructing feasible timetables. Optim. Methods Software **26**(2), 281–294 (2011). doi:10.1080/10556781003664739

[Mat68] Matula, D.W.: A min-max theorem for graphs with application to graph coloring. SIAM Rev. **10**(4), 467–490 (1968). doi:10.1137/1010115

[MBHS02] Merlot, L.T.G., Boland, N., Hughes, B.D., Stuckey, P.J.: A hybrid algorithm for the examination timetabling problem. In: Proceedings of the 4th International Conference on the Practice and Theory of Automated Timetabling (PATAT), pp. 207–231. Springer, Heidelberg (2002). doi:10.1007/978-3-540-45157-0_14

[McC07] McCollum, B.: A perspective on bridging the gap between theory and practice in university timetabling. In: Burke, E.K., Rudová, H. (eds.) Practice and Theory of Automated Timetabling VI. Lecture Notes in Computer Science, vol. 3867, pp. 3–23. Springer, Berlin/Heidelberg (2007). doi:10.1007/978-3-540-77345-0_1

[MF04] Michalewicz, Z., Fogel, D.B.: How to Solve It: Modern Heuristics. Springer, Berlin (2004)

[Mim] Mimosa Scheduling Software. http://mimosasoftware.com (2014). Accessed Jan 2014

[MM10] Müller, T., Murray, K.: Comprehensive approach to student sectioning. Ann. Oper. Res. **181**(1), 249–269 (2010). doi:10.1007/s10479-010-0735-9

[MMR07] Murray, K., Müller, T., Rudová, H.: Modeling and solution of a complex university course timetabling problem. In: Burke, E.K., Rudová, H. (eds.) Practice and Theory of Automated Timetabling VI. Lecture Notes in Computer Science, vol. 3867, pp. 189–209. Springer, Berlin/Heidelberg (2007). doi:10.1007/978-3-540-77345-0_13

[MMZ+01] Moskewicz, M.W., Madigan, C.F., Zhao, Y., Zhang, L., Malik, S.: Chaff: engineering an efficient SAT solver. In: Proceedings of the 38th Annual Design Automation Conference, pp. 530–535. ACM, New York (2001)

[Moh07] Mohar, B.: Kempe equivalence of colorings. In: Bondy, A., Fonlupt, J., Fouquet, J.-L., Fournier, J.-C., Ramirez Alfonsin, J.L. (eds.) Graph Theory in Paris. Trends in Mathematics, pp. 287–297. Birkhäuser, Basel (2007). doi:10.1007/978-3-7643-7400-6_22

[Moo91] Mooney, E.: Tabu search heuristics for resource scheduling. Ph.D. thesis, Purdue University (1991)

[MOS+13] Martin, S., Ouelhadj, D., Smet, P., Berghe, G.V., Özcan, E.: Cooperative search for fair nurse rosters. Expert Syst. Appl. **40**(16), 6674–6683 (2013). doi:10.1016/j.eswa.2013.06.019

[MPMÖ13] Muklason, A., Parkes, A.J., McCollum, B., Özcan, E.: Initial results on fairness in examination timetabling. In: Proceedings of 6th Multidisciplinary International Conference on Scheduling: Theory and Applications (MISTA), pp. 777–780, (2013)

[MR12] Müller, T., Rudová, H.: Real-life curriculum-based timetabling. In: Kjenstad, D., Riise, A., Nordlander, T.E., McCollum, B., Burke, E. (eds.) Proceedings of the 9th International Conference on the Practice and Theory of Automated Timetabling (PATAT), pp. 57–72 (2012)

[MSP+10] McCollum, B., Schaerf, A., Paechter, B., McMullan, P., Lewis, R., Parkes, A.J., Gaspero, L.D., Qu, R., Burke, E.K.: Setting the research agenda in automated timetabling: the second international timetabling competition. INFORMS J. Comput. **22**(1), 120–130 (2010)

[Mül09] Müller, T.: ITC2007 solver description: a hybrid approach. Ann. Oper. Res. **172**(1), 429–446 (2009). doi:10.1007/s10479-009-0644-y

[NMCR12] Nothegger, C., Mayer, A., Chwatal, A., Raidl, G.R.: Solving the post enrolment course timetabling problem by ant colony optimization. Ann. Oper. Res. **194**(1), 325–339 (2012). doi:10.1007/s10479-012-1078-5

[Ogr10] Ogryczak, W.: Bicriteria models for fair and efficient resource allocation. In: Proceedings of 2nd International Conference on Social Informatics (SocInfo), pp. 140–159 (2010). doi:10.1007/978-3-642-16567-2_11

[Öst02] Östergård, P.R.J.: A fast algorithm for the maximum clique problem. Discrete Appl. Math. **120**(1), 197–207 (2002)

[Pen07] Pentico, D.W.: Assignment problems: a golden anniversary survey. Eur. J. Oper. Res. **176**(2), 774–793 (2007). doi:10.1016/j.ejor.2005.09.014

[PGRD] Paechter, B., Gambardella, L.M., Rossi-Doria, O.: International timetabling competition 2002. http://www.idsia.ch/Files/ttcomp2002 (2013). Accessed Oct 2013

[Pil13] Pillay, N.: A survey of school timetabling research. Ann. Oper. Res. 1–33 (2013). doi:10.1007/s10479-013-1321-8

[PS98] Papadimitriou, C.H., Steiglitz, K.: Combinatorial Optimization; Algorithms and Complexity. Dover Publications, New York (1998)

[PZ11] Punnen, A.P., Zhang, R.: Quadratic bottleneck problems. Nav. Res. Logist. **58**(2), 153–164 (2011). doi:10.1002/nav.20446

[QB08] Qu, R., Burke, E.K.: Hybridizations within a graph-based hyper-heuristic framework for university timetabling problems. J. Oper. Res. Soc. **60**(9), 1273–1285 (2008)

[QBM+09] Qu, R., Burke, E.K., McCollum, B., Merlot, L.T.G., Lee, S.Y.: A survey of search methodologies and automated system development for examination timetabling. J. Sched. **12**(1), 55–89 (2009). doi:10.1007/s10951-008-0077-5

[R D08] R Development Core Team. R: A Language and Environment for Statistical Computing. R Foundation for Statistical Computing, Vienna (2008). ISBN 3-900051-07-0

[Raw99] Rawls, J.: A Theory of Justice, revised edition. Belknap Press of Harvard University Press, Cambridge (1999)

[Reg12] Regierung der Bundesrepublik Deutschland. Vierter Bericht über die Umsetzung des Bologna-Prozesses in Deutschland. Drucksache 17/8640 (2012). http://www.bmbf.de/pubRD/umsetzung_bologna_prozess_2012.pdf. Accessed Feb 2012

[RM03] Rudová, H., Murray, K.: University course timetabling with soft constraints. In: Burke, E., Causmaecker, P.D. (eds.) Practice and Theory of Automated Timetabling, Selected Papers. Lecture Notes in Computer Science. Springer, Berlin (2003)

[RMM11] Rudová, H., Müller, T., Murray, K.: Complex university course timetabling. J. Sched. **14**(2), 187–207 (2011). doi:10.1007/s10951-010-0171-3

[RTL76] Rose, D.J., Tarjan, R.E., Lueker, G.S.: Algorithmic aspects of vertex elimination on graphs. SIAM J. Comput. **5**(2), 266–283 (1976). doi:10.1137/0205021

[SA98] Sokkalingam, P.T., Aneja, Y.P.: Lexicographic bottleneck combinatorial problems. Oper. Res. Lett. **23**(1–2), 27–33 (1998). doi:10.1016/S0167-6377(98)00028-5

[Sch99] Schaerf, A.: A survey of automated timetabling. Artif. Intell. Rev. **13**(2), 87–127 (1999). doi:10.1023/A:1006576209967

[SH07] Schimmelpfeng, K., Helber, S.: Application of a real-world university-course timetabling model solved by integer programming. OR Spectrum **29**(4), 783–803 (2007)

[Sin05] Sinz, C.: Towards an optimal CNF encoding of boolean cardinality constraints. In: Proceedings of the 11th International Conference on Principles and Practice of Constraint Programming (CP 2005), Sitges, pp. 827–831 (2005)

[SK08] Soomer, M.J., Koole, G.M.: Fairness in the aircraft landing problem. In: Proceedings of the Anna Valicek Competition (2008)

[SMO+12] Smet, P., Martin, S., Ouelhadj, D., Özcan, E., Berghe, G.V.: Investigation of fairness measures for nurse rostering. In: Proceedings of the 9th International

Conference on the Practice and Theory of Automated Timetabling (PATAT), pp. 369–372 (2012)

[SPMFOB13] Sánchez-Partida, D., Martínez-Flores, J.L., Olivares-Benítez, E.: Modeling and solving a timetabling problem considering time windows and consecutive periods. In: Proceedings of the 5th International Conference on Applied Operational Research (2013)

[SW68] Szekeres, G., Wilf, H.S.: An inequality for the chromatic number of a graph. J. Combin. Theory **4**(1), 1–3 (1968). doi:10.1016/S0021-9800(68)80081-X

[Syl] Scientia Syllabus Plus Enterprise Foundation. http://www.scientia.com (2014). Accessed Jan 2014

[TBM07] Tuga, M., Berretta, R., Mendes, A.: A hybrid simulated annealing with Kempe chain neighborhood for the university timetabling problem. In: Proceedings 6th ACIS International Conference on Computer and Information Science (ACIS-ICIS), pp. 400–405 (2007)

[TD96] Thompson, J.M., Dowsland, K.A.: General cooling schedules for a simulated annealing based timetabling system. In: Proceedings 1st International Conference on the Practice and Theory of Automated Timetabling (PATAT), pp. 345–363 (1996). doi:10.1007/3-540-61794-9_70

[TD98] Thompson, J.M., Dowsland, K.A.: A robust simulated annealing based examination timetabling system. Comput. Oper. Res. **25**(7–8), 637–648 (1998). doi:10.1016/S0305-0548(97)00101-9

[Ueh02] Uehara, R.: Linear time algorithms on chordal bipartite and strongly chordal graphs. In: Widmayer, P., Eidenbenz, S., Triguero, F., Morales, R., Conejo, R., Hennessy, M. (eds.) Automata, Languages and Programming. Lecture Notes in Computer Science, vol. 2380, pp. 993–1004. Springer, Berlin/Heidelberg (2002). doi:10.1007/3-540-45465-9_85

[UK11] Uchida, M., Kurose, J.: An information-theoretic characterization of weighted α-proportional fairness in network resource allocation. Inform. Sci. **181**(18), 4009–4023 (2011). doi:10.1016/j.ins.2011.05.001

[Uni] UniTime: A comprehensive university timetabling system. http://www.unitime.org (2013). Accessed Nov 2013

[VG08] Van Gelder, A.: Another look at graph coloring via propositional satisfiability. Discrete Appl. Math. **156**(2), 230–243 (2008)

[Wan06] Wanka, R.: Approximationsalgorithmen: Eine Einführung. Leitfäden der Informatik. B. G. Teubner GmbH (2006). doi:10.1007/978-3-8351-9067-2

[Wei02] Weicker, K.: Evolutionäre Algorithmen. Leitfäden der Informatik. Teubner, Stuttgart (2002)

[Wil45] Wilcoxon, F.: Individual comparisons by ranking methods. Biometrics Bull. **1**(6), 80–83 (1945). doi:10.2307/3001968

[WP67] Welsh, D.J.A., Powell, M.B.: An upper bound for the chromatic number of a graph and its application to timetabling problems. Comput. J. **10**(1), 85–86 (1967). doi:10.1093/comjnl/10.1.85

[Yag88] Yager, R.R.: On ordered weighted averaging aggregation operators in multicriteria decisionmaking. IEEE Trans. Syst. Man Cybern. **18**(1), 183–190 (1988). doi:10.1109/21.87068

[Zuc07] Zuckerman, D.: Linear degree extractors and the inapproximability of max clique and chromatic number. Theory Comput. **3**(6), 103–128 (2007). doi:10.4086/toc.2007.v003a006

Author's Own Publications

[HMHW11*] Hoffmann, M., Mühlenthaler, M., Helwig, S., Wanka, R.: Discrete particle swarm optimization for TSP: theoretical results and experimental evaluations. In: Proceedings of International Conference on Adaptive and Intelligent Systems (ICAIS), pp. 416–427 (2011)

[MW10a*] Mühlenthaler, M., Wanka, R.: Improving bitonic sorting by wire elimination. In: Proceedings 23rd PARS-Workshop on Parallel Systems and Architectures of the 23rd International Conference on Architecture of Computing Systems (ARCS), pp. 15–22 (2010)

[MW10b*] Mühlenthaler, M., Wanka, R.: A novel event insertion heuristic for finding feasible solutions of course timetabling problems. In: Proceedings of 8th International Conferene on the Practice and Theory of Automated Timetabling (PATAT), pp. 294–304 (2010)

[MW12*] Mühlenthaler, M., Wanka, R.: Fairness in academic course timetabling. In: Proceedings of 9th International Conference on the Practice and Theory of Automated Timetabling (PATAT), pp. 114–130 (2012)

[MW13*] Mühlenthaler, M., Wanka, R.: A decomposition of the max-min fair curriculum-based course timetabling problem. In: Proceedings of the 6th Multidisciplinary International Conference on Scheduling: Theory and Applications (MISTA), pp. 300–313 (2013)

[OHMW11*] Omeltschuk, L., Helwig, S., Mühlenthaler, M., Wanka, R.: Heterogeneous constraint handling for particle swarm optimization. In: Proceedings of IEEE Swarm Intelligence Symposium (SIS) (2011)

M. Mühlenthaler, *Fairness in Academic Course Timetabling*, Lecture Notes in Economics and Mathematical Systems 678, DOI 10.1007/978-3-319-12799-6

Index

M. Mühlenthaler, *Fairness in Academic Course Timetabling*, Lecture Notes in Economics and Mathematical Systems 678, DOI 10.1007/978-3-319-12799-6

MIX
Papier aus verantwortungsvollen Quellen
Paper from responsible sources
FSC® C105338

If you have any concerns about our products,
you can contact us on
ProductSafety@springernature.com

In case Publisher is established outside the EU,
the EU authorized representative is:
Springer Nature Customer Service Center GmbH
Europaplatz 3, 69115 Heidelberg, Germany

Printed by Libri Plureos GmbH
in Hamburg, Germany